D1159636

Great Flower Books
1700–1900

Double White Camellia
Double Striped Camellia
Clara Maria Pope del.
Weddell sculp.

Great Flower Books
1700–1900

A Bibliographical Record of Two Centuries of Finely-Illustrated Flower Books

By

SACHEVERELL SITWELL

and

WILFRID BLUNT

THE BIBLIOGRAPHY EDITED BY
Patrick M. Synge

AND COMPILED BY
W. T. Stearn, Sabine Wilson and Handasyde Buchanan

WITH A FOREWORD BY
S. Dillon Ripley

H. F. & G. WITHERBY LTD

First published in Great Britain 1956 by William Collins

This edition first published in Great Britain 1990 by
H. F. & G. WITHERBY LTD
14 Henrietta Street, London WC2E 8QJ

ISBN 0 85493 202 X

Design by Ann Harakawa Inc.

Printed in Japan

GREAT FLOWER BOOKS: 1700–1900
was designed by Ann Harakawa and Hideshi Fujimaki, New York.
The text type is Bembo set on the Mergenthaler Linotron by
Trufont Typographers, Inc., Hicksville, New York.
The four-color separations, printing, and binding were done by
Dai Nippon Printing Company Limited, Tokyo, Japan.
The paper is 106-pound Espel Matte.

PLATE 1 [FRONTISPIECE]
Clara Maria Pope drew the five folio plates
of nurseryman Samuel Curtis's A Monograph on the Genus Camellia *(London 1819).*
Weddell's aquatint etching of 'Double White Camellia'
and 'Double Striped Camellia' was colored by the artist in gouache.
The glossiness of the leaves is conveyed with gum arabic,
a clear and shiny varnish used often by print colorists.

PLATE 2
Maria Sibylla Merian produced a book on European insects
in 1679 before treating the denizens of the South American colony of Surinam in
Metamorphosis insectorum surinamensium *(Amsterdam 1705).*
Sluyter has engraved the banana flower and its attendant insects with a bold line,
and this print has equally bold hand-coloring.

P. Sluyter Sculp
12

CONTENTS

NOTE TO THE NEW EDITION

THIS VOLUME GUIDES US THROUGH an extraordinary body of art and science. It documents a period when these two strands of culture were tightly interwoven; a time when the natural sciences spawned the most artful expression of the printer's craft ever achieved. Through its superb visual record and its comprehensive bibliographic catalogue, *Great Flower Books: 1700–1900* provides collectors with an unparalleled introduction to the great natural history books and prints.

The Bibliography, not limited to cataloguing a single library or collection, nor being a strictly scientific compilation, encompasses the breadth of two centuries of exquisite colorplate books with the collector in mind. The field is not confined to the giants of the era, such as Pierre Joseph Redouté and Dr. Robert Thornton; it includes more obscure but no less superb works such as Samuel Curtis's *The Beauties of Flora* (plates 31 and 32), of which very few copies are known to exist. The Index leads the reader to the scattered contributions of individual authors, artists, and printmakers, from the rare excellence of Gerrit van Spaendonck to the accomplished prolificacy of Walter Hood Fitch. Thus, *Great Flower Books* is perhaps the greatest single guide to these prints, allowing collectors, both avid and casual, to become better acquainted with the books and prints they own or would like to own.

We have added to Sacheverell Sitwell's original text Handasyde Buchanan's Appendix to define the language of connoisseurship, which allows a collector to know and discuss prints with precision. We are grateful to publisher Weidenfeld and Nicholson for permission to reprint this section as well as Buchanan's enlightening Introduction to his own book, *Nature into Art*. As a reflection of the increasing interest in the field, we have been able to greatly increase the number of titles in the reference bibliography with books published since *Great Flower Books* appeared in 1956.

We were prompted to republish this volume because, although *Great Flower Books* is quoted extensively in the catalogues of auction houses, dealers, and museums, it is largely unavailable to the private collector. The large size of the original edition (14″ by 20″) confines its use to the floor or the dining table, and libraries often limit its circulation to a rare book reading room. From time to time the original edition (limited to 2,000 copies) appears in a book dealer's catalogue for around $1,000.

The editors are pleased to make Sitwell's and Wilfrid Blunt's texts, along with the scholarly bibliographic work of W. T. Stearn, Sabine Wilson, and Buchanan, widely available in a convenient and well-illustrated format. *Great Flower Books* is a unique authority. It combines the expertise of naturalist and man of letters Sacheverell Sitwell, art historian Wilfrid Blunt, botanist W. T. Stearn, and London book dealer Handasyde Buchanan. This constellation results, perhaps, in a certain anglophilia, but those familiar with the original edition will note that we have illustrated works from a greater diversity of sources, to more accurately reflect the overwhelming contributions of European artists and printmakers. Indeed, many of the greatest engravers and printers in London arrived there from the continent with their techniques fully mastered.

Revisions to the text are limited to the clarification of Sitwell's sometimes casual references to authors and titles. The List of the Plates offers complete descriptions of the fifty-two prints reproduced. A look through the list reveals the variety of collections from which we have photographed fine examples of colorplates for our illustrations.

The editors wish to thank all those who have helped prepare this book, but particularly the librarians, curators, and rare-book dealers who made their collections, time, and support so readily available. Graham Arader and Donald Heald, renowned New York dealers of fine colorplate books, opened their galleries to us and gave valuable counsel on our selection of images. The New York Public Library was an extraordinarily rich source of material and assistance, and we especially thank Bernard McTigue, curator of the Arents Collection, and John Rathe of the Rare Book Room for their constant cooperation and support. We are also indebted to Charlotte Tancin, librarian of the Hunt Institute and provider of counsel beyond the Institute's representation in this volume. The assistance provided by the staff of the Library of the New York Botanical Garden, in particular by research librarian Bernadette Callery, Lothian Linus, and Judy Reed, was indispensable. Hirschl & Adler Gallery gave us access to their excellent file of transparencies for Allen's *Victoria regia*. Thanks also to Melissa Moore and Dennis Jon of the Minneapolis Institute of Arts, Linda Lott of Dumbarton Oaks Garden Library, and Stanley Crane of Pequot Library. We also appreciate the interest and research afforded us by Cornell University and Yale University, whose libraries we did not ultimately call upon. We are grateful to Robert Lorenzson, who photographed most of the prints, for his superb skill and keen appreciation of the subject.

Finally, the editors wish to thank Elizabeth Braun, without whose expertise, inspiration, and unflagging energy this book could not have been produced. Her love of fine colorplate books and her attention to detail are evident on every page of this volume. She was tireless in her search for the best possible examples of each plate we have illustrated, and her captions and List of the Plates attest to her exacting scholarship. She also took pains to emend the text and bibliography where errors or the scholarship of the past thirty years called for it.

THE EDITORS

FOREWORD

IN THIS REMARKABLE PUBLICATION cataloguing two centuries of plant and flower printmaking, Sacheverell Sitwell and Wilfrid Blunt have created an exceptional work in text and illustration. This volume is a veritable floral garland.

Walking into a library in the British Museum on a dull, dark day, Sitwell writes of coming upon a brilliant medieval illuminated manuscript. The scribe of such a manuscript could well have been a monk, removed from the collecting of plants far afield, and confined to the flora in the monastery garden. Yet he made his own tools, ground his own pigments, applied and burnished the gold leaf as if recording the rarest, most ephemeral blooms. Thus does Sitwell introduce us to a visual feast, for the old pictures of flowers and plants are scrumptious. They make one feel hungry to touch, smell, and taste them, as we do with leaves in the herb garden.

Painters of flowers are perhaps more numerous than painters of birds. Certainly, their folios of original paintings and drawings are in every great herbarium, such as Kew in England, the queen of the world's herbaria. Much of the identification of plants today derives from original paintings and the handwritten descriptions that go with them. It is still not uncommon for a botanical expedition to have a plant artist among the company so that flowers collected in an inaccessible region can be illustrated at once.

One such twentieth-century illustrator died only recently. Margaret Mee's watercolors are in herbaria all across the United States and Great Britain. Her work, stretching deep into the wild Amazon basin, is exceptional and quite in the tradition of the great eighteenth-century books and watercolors hidden away in libraries and museums. Another prominent woman artist who specialized in tropical flora is Maria Sibylla Merian (plates 2 and 3). She journeyed to Surinam, not far from Margaret Mee's collecting ground, at the end of the seventeenth century, and with a pioneer's courage spent two years in the wilderness. Her drawings are splendid, embodying the natural history of butterflies and moths as well as plants, and illustrating a rare and curious voyage.

Great Flower Books offers a selection of prints by the most celebrated of all the plant artists, Pierre Joseph Redouté, whose inimitable pastel colors are reproduced here with striking faithfulness. Redouté's popularity has been overwhelming and his renown has lived on in his timeless paintings and prints of roses and lilies. But as the authors point out, there were many other artists at work in the early nineteenth century. Although less well known today, their work meets the standard set by Redouté. Sitwell and Blunt mention Turpin and van Spaendonck, and their illustrations, like those of Prévost and Bessa, deserve far greater celebrity than they currently receive.

At the end of the nineteenth century we still see magnificent publications brought about through the impetus of botanical exploration, which reached its peak just before the turn of the century. One can think of Sir Joseph Hooker, whose productivity in delineating the flora of the Himalayas was so great. I remember very well some of the British botanical explorers I encountered during my own travels in the Himalayas, and I am sorry that there are not more of them working today. I can think of "Chinese" Wilson and Frank Kingdon-Ward in my time, part of a continuum of explorers in the Old World. They are currently matched in the New World by David Fairchild and Wilson Popenoe, and on to those working with the New York Botanical Garden, the Smithsonian, and other likeminded institutions. What enthralling careers these adventurers have had! One's nostalgia for those days becomes patent in travel itself and in reading about it.

What a pleasure it is to find flower books so thoroughly collated by Patrick Synge, a great author in his field, with W. T. Stearn, the paramount authority on botanical history, as well as Handasyde Buchanan, botanical bibliographer and book collector. I can only hope that today's writers and explorers will continue the tradition of botanical books begun in the medieval monasteries so many hundreds of years ago.

There is something particularly captivating about botanical books to anyone who appreciates flower painting, for they comprise an ever-unfolding treasure house. And the delight found in turning pages never ends, just as the best nursery catalogues of plants for sale hypnotize us each spring. The portrayals of flowers in these great books, so accurate and beautiful, are visions of plants that strive to satisfy and soothe the mind and spirit.

There is a symmetry in the arrangement of this volume and its companion, *Fine Bird Books: 1700–1900*, published simultaneously by Atlantic Monthly Press, which is wholly appropriate. Just as birds have highly attuned color perception, so we humans appreciate the colors of plants and birds blended together. We undervalue the worth of beautiful illustrated volumes today, so rapidly and profoundly has technology transformed our existence. These two books are an entrance into another world which we are in peril of losing, and they serve to remind us of its lasting fascination.

Bibliographies of colorplate natural history books are of extreme importance, for a catalogue of the greatest books allows one to browse through the antique realm they represent. The old books are an invaluable resource to the scientist and the collector, since much of our present knowledge of science originated from these sources.

As the world changes—its variety of species and even its wildernesses contracting—we must be thankful that so much has been done to document our resources and surroundings. Without such collections of volumes it would be almost impossible to reconstruct the origins of plants and animals and their milieu. One wonders what the future holds for so much of this glory. Let the flowers pictured in these pages remind us of our responsibilities to preserve their world and the books that record it.

S. Dillon Ripley

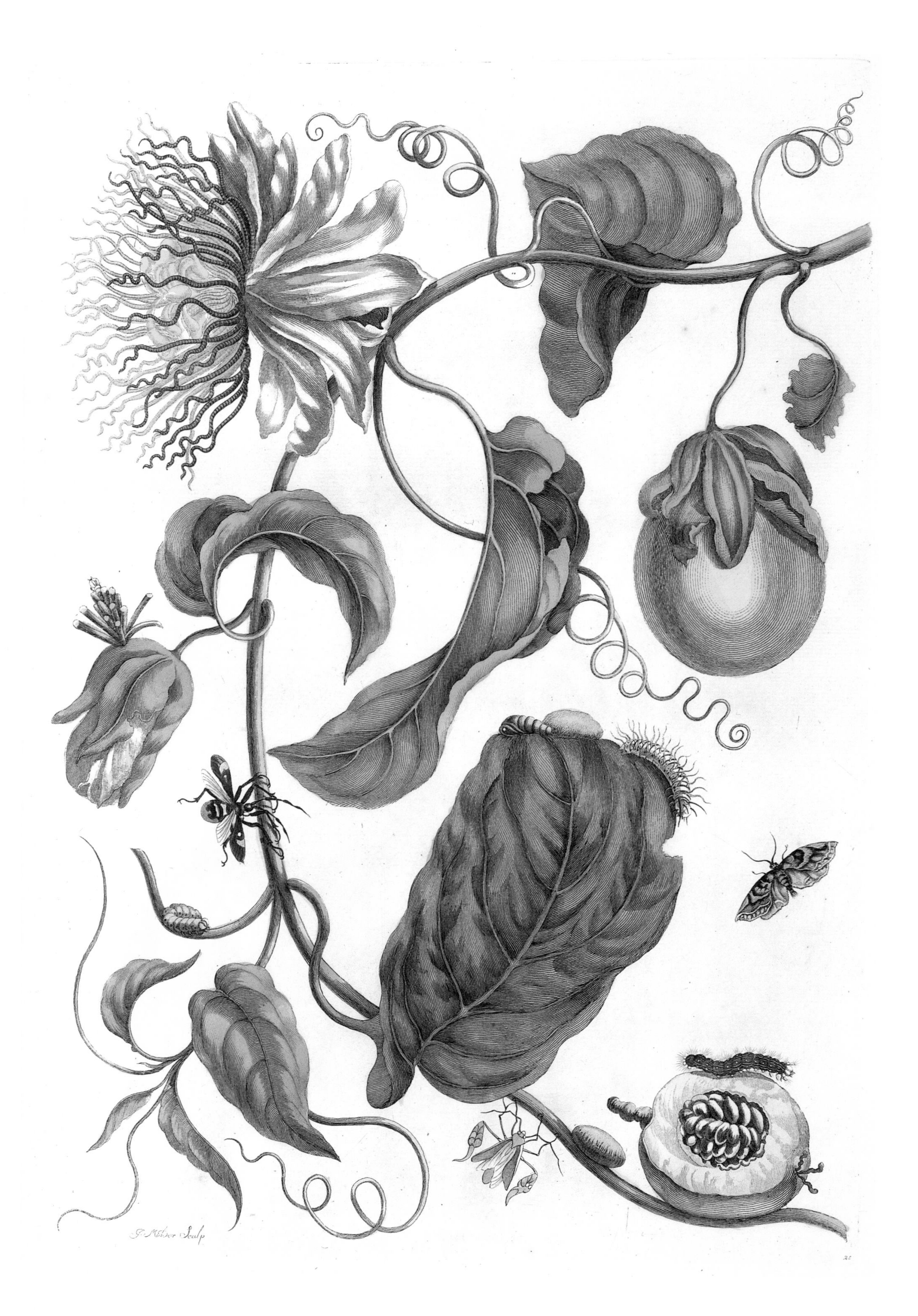
J. Mulder Sculp
21

INTRODUCTION

Handasyde Buchanan

THE SCOPE OF THIS BOOK is the period from about 1700 to 1900 or a little earlier, and its aim is to show a selection of the best illustrations from old natural history books with coloured plates, together with an account of these books and their creators, which represent what might be called the golden age of the natural history book. It is not intended to be a complete history of books in this period, but rather a personal choice from among the best examples. The majority of these books were about birds and flowers, although insects are, indirectly, well represented since almost all books about them are, from the point of view of their plates, in effect flower books.

At the beginning of our period—the time when books with coloured plates were first produced—the pictures were taken from copper-plate engravings, and coloured by hand. This method was used from 1700 until about 1830 when that type of illustration was outmatched by the aquatint, the mezzotint and the stipple engraving, often printed in colours. This was the finest hour of colour printing. Around 1830 the lithograph appeared, which involved printing from a stone instead of a copper plate; it was still, however, coloured by hand. But towards the end of the nineteenth century the chromolithograph, in which the colour came directly from the stone and was not applied by hand, had won the battle. Of course it made natural history books grow steadily cheaper, and was the parent of all later flower, bird and animal plates, but these illustrations lack the individual artistry that made the earlier ones fascinating and unique, and the pleasure of looking at the pictures as works of art has gone. An account of the differences between these various processes, and how each was done, appears at the end of the book, on page 175. However in the 1950s a number of superbly produced, definitive books were published, all with bibliographies and fine reproductions of plates, such as *Great Flower Books* and *Fine Bird Books.* I contributed to most of them. Now these books have themselves become treasured, collectable items.

During the eighteenth and nineteenth centuries certain countries were at the top of what might be called the natural history book league, and this is perhaps the place to say who they were, when and why. The Dutch came first because they were the earliest explorers—Tasmania, for instance, was originally discovered by the Dutch in the seventeenth century, and known as Van Diemen's Land. Maria Sibylla Merian's *Insects of Surinam* (Surinam being a Dutch colony in northern South America at that time) of 1705 was the best example of Dutch work, although as late as 1794 a lovely and anonymous flower book called *Nederlandsch Bloemwerk* showed that the Dutch had not forgotten how to produce

PLATE 3

Maria Sibylla Merian's drawing of a passionflower was robustly engraved by Joseph Mulder to illustrate Metamorphosis insectorum surinamensium *(Amsterdam 1705). The artist traveled to the Dutch colony of Surinam with her daughter Dorothea in the 1690s, and systematically recorded the metamorphosis of insects.*

Helleborus niger Officinarum
Ellebore noir usuel .
N. Robert del. et sc.

beautiful work. The Germans were active in the early eighteenth century, and J. W. Weinmann's *Phytanthoza Iconographia*, a mammoth book in folio format, comprising four volumes with 1,026 plates, is perhaps the most notable example. But from 1730 onwards the British really dominated the field with a very large number of great books.

There is, of course, one colossal exception in these years. France, with Redouté for flowers and Levaillant for birds, using stipple engravings expertly printed in colour, undoubtedly created the finest books of all between 1790 and 1830, despite competition from excellent works such as those of Thornton and Brookshaw in England. The United States—although of course it produced the magnificent Audubon—does not really enter into this particular discussion, since it must be remembered that Audubon's gigantic masterpiece, *The Birds of America*, was actually printed and published in London.

One interesting and sometimes confusing point is that some famous books are known by the author's name, and not by that of the artist. Redouté, the greatest of all French flower book artists, has his name on almost all books with which he was associated—Redouté's *Roses*, Redouté's *Liliacées*, and so on—though not always as the principal. Levaillant, however, whose name appears on most of the great French bird books of the same period, was a naturalist who travelled far, but never to my knowledge drew any of the pictures for his books—almost all were drawn by Barraband. Dr Thornton was the author of perhaps the most splendid of all English flower books, popularly known as *The Temple of Flora*, but whose real title is *New Illustration of the Sexual System of Linnaeus*. Of the justly famous plates which illustrate this book only one, the Roses, was drawn by Thornton himself. Books are sometimes known by the artist's name, and sometimes by the name of the author of whatever text they may contain. A number of different activities were involved in producing a book, and although some tasks were often performed by the same person, it was quite possible for an explorer to provide the specimens, which would then be written up by someone else, illustrated by a professional artist, engraved by another hand, and perhaps hand-coloured by a whole team of people. All this explains the plethora of credits which are often to be found at the foot of natural history plates. The abbreviations used by artists, engravers and printers are given on page 179.

Although this book is only intended to cover printed books with coloured plates, it is worth saying a word or two about the original watercolours. Some artists were admirably served by their engravers and printers—despite Audubon's immense talent as a bird artist, for instance, *The Birds of America* would be less of a masterpiece were it not for the skills of Robert Havell. Other artists, however, were treated less well than they deserved, and so, perhaps unjustifiably from the artist's point of view, their books are considered less good. But all natural history artists are best judged by their original

PLATE 4

This uncolored print exemplifies the brilliance that botanical engraving achieved in the hands of master printmakers. "Ellebore noir usuel" illustrates Nicolas Robert's Estampes pour servir à l'histoire des plantes *(Paris 1701), the herald of our era of great flower books. The interlacing tapered lines describe color, anatomical details, and the illusion of three dimensions.*

work, and anyone who has the opportunity of seeing the watercolours for *The Birds of America* at the New York Historical Society, or Georg Ehret's *vélins* (paintings on vellum) at the Victoria and Albert Museum and the Library of the Royal Botanic Gardens, Kew, will be well rewarded for the effort.

The part played by the great Swedish naturalist Linnaeus (1707–78) cannot be underestimated. He made an enormous contribution to natural history by tabulating a completely new system of classification for plants and animals. Before Linnaeus classification had been a somewhat haphazard affair and lacked any sort of standardization. The most commonly followed system for plants had been that of Tournefort, who based his distinctions on the shape of the corolla. Linnaeus, on the other hand, defined his categories by the sexuality of plants—a system that scandalized the prurient morality of the Victorians, who felt that young girls in particular should not be exposed to such grossness! In his *Species Plantarum* of 1753, and *Genera Plantarum* (1737; fifth and most important edition, 1754), Linnaeus divided plants, after examination of their male sexual organs, into 24 classes, which in turn he subdivided into orders *vis-à-vis* their female sexual organs. The names were derived from the Greek—lilies, for instance (as illustrated in Mrs Bury's *Hexandrian Plants*), which have six stamens, are in the class 'Hexandria' (from the Greek words for 'six' and 'male') and of the order 'Monogynia' (from the Greek for 'one' and 'female'), since they have only one style. Linnaeus was an imaginative man and did not lack a sense of humour: he described the class Polyandria (many male organs), which includes the poppy, as 'Twenty males or more in the same bed with the female'.

His new system of classification for animals was far less revolutionary than that for plants. The first edition of the *Systema Naturae*, published in Leyden in 1735, divided them roughly into quadrupeds, birds, amphibians, fish, insects and invertebrates (*Quadrupedia*, *Aves*, *Amphibia*, *Pisces*, *Insecta* and *Vermes*). In the tenth edition of 1758 he subdivided these classes, as with plants, and by this time he had recognized the importance of differentiating between mammals and other animals, thus *Quadrupedia* became *Mammalia* (and included man). But his system was not universally welcomed; Thomas Pennant, the English naturalist, wrote in his *History of Quadrupeds* of 1781—perhaps foreshadowing the furore over Darwin's theories of evolution a century later: 'My vanity will not suffer me to rank mankind with Apes, Monkies, Maucaucos and Bats.'

While this book is concerned with the art and beauty of the natural history book rather than with its botanical or zoological merits, a number of expeditions—and subsequently books—were undoubtedly embarked on as a direct result of the availability of this relatively simple and foolproof system of classification, and lovers of old natural history books have good cause to be grateful to Linnaeus.

We have, of course, reached a stage today when no one but a fool would bet on the value of any book, because of inflation. But it is worth having a look at the past in this connection. *The Temple of Flora*, which listed in 1890 catalogues at £20 (!), was on sale for £100 when I started as a bookseller in 1930. When I returned after the Second World War and found that the price of everything had escalated, it was then £500. By 1965 (it is now very scarce in the right state and in good condition) that price had shot up to £4,000–5,000. It may now be £12,000 or more; good copies rarely, if ever, turn up.†

The value of a book depends on so many different reasons—date, size (book sizes are dealt with on page 179), number of volumes and plates, intrinsic beauty, scientific merits (this is perhaps less important than it should be, and someone from a more scientific background than I should say whether this is true or not); but ultimately on its scarcity. There are from six to ten copies of Samuel Curtis's *The Beauties of Flora* in existence, and indeed it is likely that no more were printed. This folio volume, published between 1806 and 1820, has only ten plates. It is in effect priceless.

Not all the books described or illustrated here really merit the epithet 'great'. But they are all of the highest quality—in other words, their printed plates are among the best ever produced. The texts in these books are frequently poor, even non-existent, or merely what we would now call extended captions. Many of them were produced over a long period, as single plates, and were thus clearly meant to form picture books or to be framed—either for their decorative qualities, or for their informative value, as when describing a newly discovered animal or plant. All the plates reproduced in this book are intended to be enjoyed as things of beauty.

† In the fall of 1988, an exquisite copy of Thornton's *Temple of Flora* sold in New York City for more than $200,000 (roughly £95,000).

1 Keysers Jewel Hyacinth.
2 Diamend d°.
3 Double blossom'd Peach.
4 Single Orange Narcissus.
5 Double Endroit Tulip.
6 Glory of ye East Auricula.
7 Double Wall flower.
8 Blush red lilly of ye Vally.
9 British King Anemone.
10 Cœlestis Anemone.
11 Amaranthus trachee.
12 Single Junquill.
13 Loves Master Auricula.
14 Double painted Lady Auricula.
15 Palurus Christs thorn.
16 White Lilly of the Vally.
APRIL
17 Merveille du Monde Auricula.
18 Lady Margareta Anemone.
19 Juliana d°.
20 Double Junquill.
21 Duke of Beauford Auricula
22 Lecreep N° 1 Tulip.
23 Beau Regard Tulip.
24 Dwarf Single flowering-Almond.
25 Duke of St. Albans Auricula
26 Turky ranunculus sweet scented.
27 Double Cuckow Flower
28 Grand Presence Auricula
29 Sea Pink.
30 Double flowering Almond.
Design'd by Ptr Casteels.
From the Collection of Robt. Furber, Gardiner, at Kensington, 1730.
Engrav'd by H. Fletcher.

THE ROMANCE OF THE FLOWER BOOK

Sacheverell Sitwell

OTHERS, AS WELL AS THE WRITER, may have discovered the miracle of flower painting during childhood, when being dragged through a museum on a dull dark day. On the day I am thinking of, so long ago, we walked back across the entrance hall of the British Museum into the Grenville Library, where there was an exhibition of medieval illuminated manuscripts; and there, in a showcase (but anticipating nothing of the kind) I found the border of a missal, a golden panel or inlay, it seemed, upon the vellum, on which with exquisite freshness and delicacy were painted wild strawberries, blue periwinkles, and a red pink or clove carnation. It set me wondering. And with intervals and intermissions I have been thinking and wondering about it ever since. For such things, however little and unimportant in dimension, are among the miracles of human taste and skill.

There is the conception, not always true, in fact, for the illuminator was more often a professional painter, that the scribe or artist was a monk. Let us keep to our illusion. So he was celibate, and therefore removed from the pleasures and temptations of the world, even in that more simple age. Then, for delight in turning over in one's own mind, there is the curing and preparing of the vellum; the making of brushes of rabbit or other hair; the collecting of goose plumes where the geese moult their feathers round the monastery pond, and the sharpening of the quills; the grinding of the colours derived from all sorts of curious sources, including a blue which in the Middle Ages they got from powdered lapis lazuli; and last but not least the patient burnishing of the gold, done with an agate and brought, thereby, to its glittering metallic finish. It was borne in upon one that they had few flowers; that a pink, a rose, a periwinkle, a lily, and a poppy, were nearly all they knew in their gardens. And the wood strawberry with its flecks of gold? Did the illuminator paint it because of its simplicity growing in the grass; or was it to him a thing of luxury, and the dessert of kings and queens? For, after all, peers' coronets are composed of strawberry leaves, and this would not be the case were wild strawberries as common and as often met with as are the daisies of the field. No, quite clearly to the medieval painter the strawberry, like the carnation and the periwinkle, was a thing to wonder at. He was in a state of innocence and wonderment. And persons who are blasé and have lost their sense of wonder are not in the right mood to apprehend beauty in a humble flower.

But the answer is that humility is not the rôle of any flower at all. Not even buttercup or dandelion. The nearer you hold those to your eye the more proud they grow in their magnificence. Though the individual buttercup be but a golden item, a stitch, as it were, in a meadow of cloth-of-

PLATE 5

Flemish artist Pieter Casteels painted "April" for Robert Furber's Twelve Months of Flowers *(London 1730), and Henry Fletcher engraved the detailed plate. Furber operated a nursery at Kensington and published twelve bouquets to advertise his wide selection of flower seeds. Twenty-seven varieties from tulips to passionflowers appear in this plate alone.*

gold, it is, yet, a golden cup or grail, and the eyes dazzle on looking into its golden circumference and at the shine and glitter of its metal. And the dandelion is, as its name suggests, lion-like and something of a dandy, whatever the true etymology may be; wearing a coat of golden colour, and rayed like the sun, itself, when that darts its sunbeams through the clouds. A creature of splendour. But, then, no flower is simple. Or take the snapdragon. Pull down its lower lip or maw, which may remind one of a secret ballot-box into which to put letters or voting-papers on some ethereal election. It is the yellow or red pillar-box of King Oberon and Queen Titania, or mail-bag of 'Queen Mab'. The Italian name for snapdragon is *bocca di leone* or 'lion-mouth,' which carries the same suggestion of the shutting of a fiery trap that closes on a spring, it could be, and the playful invitation to open it again and have another look inside. Then you behold something speckled like the thrush's throat, and may marvel how so lovely a song is born of a diet of worms. And how many of such wonders to the stem! For it can be as prolific as the foxglove on a single spire.

In this same illuminated missal, or another one near by, there was a picture of a harvest field with poppies and cornflowers growing in the corn. You could hear, it seemed to me, the 'schresh' 'schresh' of the reapers as they stood to whet their scythes. And it was an intimation that all garden flowers in their origins come from the woods and fields. All are either prisoners or uninvited guests. The first pink, cousin of the 'Cheddar' pink, may have grown unsuspected out of a chink in an old wall. Collected, thence, and cared for, by a Carthusian monk in a white robe. Flower seeds may be carried home in loving hands; or loutishly upon a yokel's boots. How did the fern come to its crevice; for there is not another within sight or knowledge? The spores were wind-blown; or carried by a bird. By what feat of magic does a fairy ring of mushrooms appear upon an unmown lawn? Such are mysteries that are easy of explanation, but no less magical once they are explained. Or the deadly nightshade comes uninvited, like the wicked fairy Carabosse, and insinuates herself among the currant-bushes in a neglected corner of the kitchen garden.

One thing, at least, could be said with truth about the medieval illuminator before we part with him, that a single lifetime is not enough. He would need a hundred lives to paint the newer families of flowers. For the increase in the garden population is a phenomenon that has got out of hand. We have only to think of a painter early in the last century at work upon the rhododendrons. He would have had little else to draw but the *Ponticums*, with little variation in shape or colour of flower, and only dark or darker leaves. But it would need many years of a lifetime to paint, accurately, the mass of later importations from the far Orient, their variants and their hybrids. Or consider irises or daffodils as they were but fifty years ago. There must be, now, a hundred for every one there was before. It must, now, be a matter of selection, not of the complete whole. For that would be beyond the bounds of possibility.

PLATE 6

"Magnolia Grandiflora" conveys the powerful draftsmanship of Georg Ehret.
It illustrates Mark Catesby's The Natural History of Carolina *(London 1731–43),*
a book with 220 plates etched almost wholly by the author.
Catesby sent many specimens of magnolia to his patrons in England
while he researched the book in America in the 1720s.
Ehret painted this flower as it bloomed in London in 1737.

Merely to paint a record of the new American irises would be ten times more than any one painter could accomplish as they come out year by year. But this twentieth century, so far as flowers go, is not only an affair of gains, for there are losses, too. Whole families of florists' flowers, the auricula, the ranunculus, the pelargonium, to take the most obvious instances that come to mind, have fallen from fashion, or like the camellia are coming slowly back into favour when their ranks are thinned and even their names are lost or doubtful.

What is now a steady stream or even torrent of importations, though stemmed in certain directions, as late as Elizabethan and Jacobean times was but a slow and intermittent trickle. The 'Turkish' lilac or 'laylock tree' of Henry VIII's garden at Nonsuch was followed half-a-lifetime later by the first tulip, also from the land of turbans. It was at about this date that the great and eminent Clusius, a Fleming of Arras, began his systematic plant collecting in Spain, in Hungary and the Austrian Alps, and from Turkey and the Levant. It is the epoch of the woodcut books; of Hans Weiditz, one of the most engaging minor artists of the German Renaissance, at not many removes from Urs Graf or from Altdorfer, and who drew the illustrations for Otto Brunfel's *Herbarium Vivae Eicones* (1530); or of Leonhart Fuchs' *De Historia Stirpium* (1542) with its team of artists. From these works of instruction and delight the next stage takes us to the hand-coloured engravings of the *Hortus Floridus* by Crispin de Passe (1614), a work no longer in the tradition of the illuminated manuscript or medieval herbal. But the marvellous flower drawings of the Fleming Joris Hoefnagel are still of that older world. His *Missale Romanum* painted for Archduke Ferdinand of Tirol, nephew of Charles V, is, in fact, last of the great illuminated manuscripts and intended, as they were, for one patron and his family and friends. But the *Hortus Floridus* of Crispin de Passe is intended for a larger public, to be looked at, perhaps, on winter evenings with snow outside and in the flickering lighting of early seventeenth century rooms, with summer still far away before even the hellebore opens its green cups in February, one of the first of their spring flowers.

It would be true, though, to say that flower drawing continued through the whole of the seventeenth century as an independent and particularized form of hand-illumination. What else, indeed, is the exquisite *Guirlande de Julie* of Nicolas Robert (1614–85), most wonderful example of the *florilegium*? But such hand-painted manuscripts or books of drawings, sumptuous or delicate, are not our subject, though a volume of reproduction at least equal if not superior to this present one could be devoted to them. For we have already arrived at the great printed flower books. But locked away in museums, and to a lesser extent in private libraries, are beautiful and quite unknown albums of flower drawings that are in prison, as it were, and only visited at rare intervals by a mere handful of amateurs and students.[1] It is not necessarily the finest of the flower masters' drawings that were engraved in books; while it is also true that the original painting, perhaps on vellum, may be more delicate and of better quality and finish than the reproduction. This hiding away and seclusion of original flower drawings which is, apparently, insurmountable and an obstacle to their general appreciation which will never be overcome, can be paralleled and brought up short and sharp to our attention if we suppose for the sake of argument that in the case of early music only those pieces engraved and printed during the composer's lifetime were ever played, and the rest remained in manuscript and unperformed. A flower master of the stature of Nicolas Robert is in exactly this position. It persists through all the period of the flower books, and there is case after case of flower painters who had, comparatively, as little of their work printed while they were alive as the *XXX Essercizi per gravicembalo* and the *Forty-two suites of lessons for the harpsichord* which were all of Domenico Scarlatti's five hundred and forty, and more, keyboard sonatas published in his lifetime. Even a painter so popular and famous in his own lifetime as Georg Dionysius Ehret is imperfectly admired and estimated when known only from engravings; while Nicolas Robert remained unknown until the public, restricted in number, who are interested in such things had its glimpse of his *Guirlande de Julie* which was the revelation of the exhibition of botanical books and drawings held by the National Book League in London in 1950. This *florilegium* and collection of madrigals (for it is both), written out in the hand of the greatest calligrapher of his time, has guarded its virginity intact, and is unpublished. It is precisely, therefore, a case in point; that only the later of the flower masters are to be judged upon their engraved works alone. However fine and beautiful the hand-colouring it will not tally with the original. It has to be accepted on general principles that the flower masters were better than their flower books, and that they were in almost every instance better painters than those who painted birds.

It so happens that Nicolas Robert's engraving of the Black Hellebore, from Denis Dodart's *Nouveau Recueil des Plantes* (1701), is the earliest of our plates. This is the Christmas Rose, the familiar white flower much esteemed in old gardens because of its unconquerable optimism and insistence upon flowering in mid-winter in spite of all warnings. It has, as well, an extraordinary length of blossoming, the flower heads keeping some semblance of their youth long after daffodils and other spring flowers are dead. We are probably correct in thinking of Nicolas Robert taking this flower indoors to draw upon a cold winter day, and it is mind pictures such as this which bring him near to us although he lived as long ago as the reign of Louis XIII. For myself, whenever I think of that far-off epoch, its architecture, its poetry or the preciosity of its flowers, I have the air of Handel's *Harmonious Blacksmith* ringing in my ears, for it is Handel's setting of a song by Louis XIII, and if you listen carefully you hear the French accent on the words. This song to which a silly legend has attached itself is of a hundred years before Handel's time.

Of all the flower masters it is Nicolas Robert who should have drawn fritillaries because his touch and line are peculiarly apt for that, and the fritillary looks to be a flower of that age. Among botanical books this is, probably, the most beautiful of unaccomplished tasks. It still awaits its master; and it is true, of course, that fritillaries unknown to the seventeenth century have been collected from California, Persia, Caucasus, and the Levant, and that none but *F. Meleagris* and one or two others grew in the old French gardens.

The age of botanical exploration is at hand, and among its earliest adepts is Maria Sibylla Merian. Clusius and other predecessors in the sixteenth century had travelled widely, but they were men of learning, botanists and not painters. But Madam Merian, of a well known family of artists, had the temerity to leave home and husband and set sail for Surinam. She took her daughter with her on this

Aloe tuberosa levis.

expedition, which would be a considerable enterprise to-day. They reached Dutch Guiana, in South America, in June 1701, and stayed there for two years studying flowers and insects, which latter were her main interest, and making several hundreds of drawings upon vellum, of which seventy-two were later engraved and hand-coloured by mother and daughter for her *Metamorphosis Insectorum Surinamensium* (1705). It was Madam Merian who discovered the birth of butterflies and moths from chrysalises, and she may have considered herself a scientist and holder of advanced religious views as much as artist.[2] The drawings in her large folio are, in fact, bold in design and of masculine touch. They are masterly, and yet they lack something. Could it be an excess of masculine temperament in Madam Merian, a touch of Rosa Bonheur, of George Eliot, of Madame Curie, of Mrs Sydney Webb? In aesthetics, they belong to an age when the hand of man, or woman, could not go wrong. Large in drawing, in spite of their tropical content they rather resemble a set of designs for embroidery, and particularly, for the work known as *peinture à l'aiguille*. They are flat, that is to say, and without depth, rather as if composed out of pressed flowers, but relieved from monotony by the shapes and colourings of the insects with which they are a-crawl. All in all, looking through her *Metamorphosis* again and admiring once more her blue water lily of the Amazon, Madam Merian is at her best, prosaically, in paintings of a pineapple and of a bunch of bananas, which is but another proof of how quickly poetry can be turned to dross and the fruits of the Hesperides become bargains at the fruit-barrow and litter of the seashore and the slippery pavements.

It should be stressed again that in order to appreciate Madam Merian at her worth it is essential to see original paintings by her upon vellum. But, in her case, somewhat of a contradiction lies herein. For, already, in her time flower painting as distinct from flower books had divided into two currents. The Dutch and Flemish flower masters, Jan Brueghel, van Aelst, van Walscapelle, de Heem, and in a later generation, Rachel Ruysch and Jan van Huysum, to name but a few out of so many, are to be altogether separated in their aim from the draughtsmen who drew on vellum or paper, who were, in fact, illustrators of works on botany. They are to be admired more than anywhere else in the magnificent *vélins* of the Jardin des Plantes in Paris, and of course their drawings exist in thousands in libraries and in private collections. But Madam Merian in her art pertains to the former school, although she is a member and ornament of the latter. In her 'style' and her 'arrangement' she is of the great Dutchmen. And among lovers of botanical drawings there is always criticism of the Dutch flower painters. They sacrificed everything to effect; and in simple words, what did they do during the rest of the year if they did not paint flowers during the six months when they were not in season? They cannot have remained idle. So they worked from notes, or after sufficient experience went straight ahead as if the flowers were there in front of them. We have only to look at the great flower pieces to know that it is

PLATE 7

Pharmacist Johann Weinmann identifies 4,000 plants in the 1,025 engravings of Phytanthoza iconographia *(Regensburg 1735–45). "Aloe tuberosa levis" is a color-printed mezzotint that captures the distinctive volume of this succulent plant.*

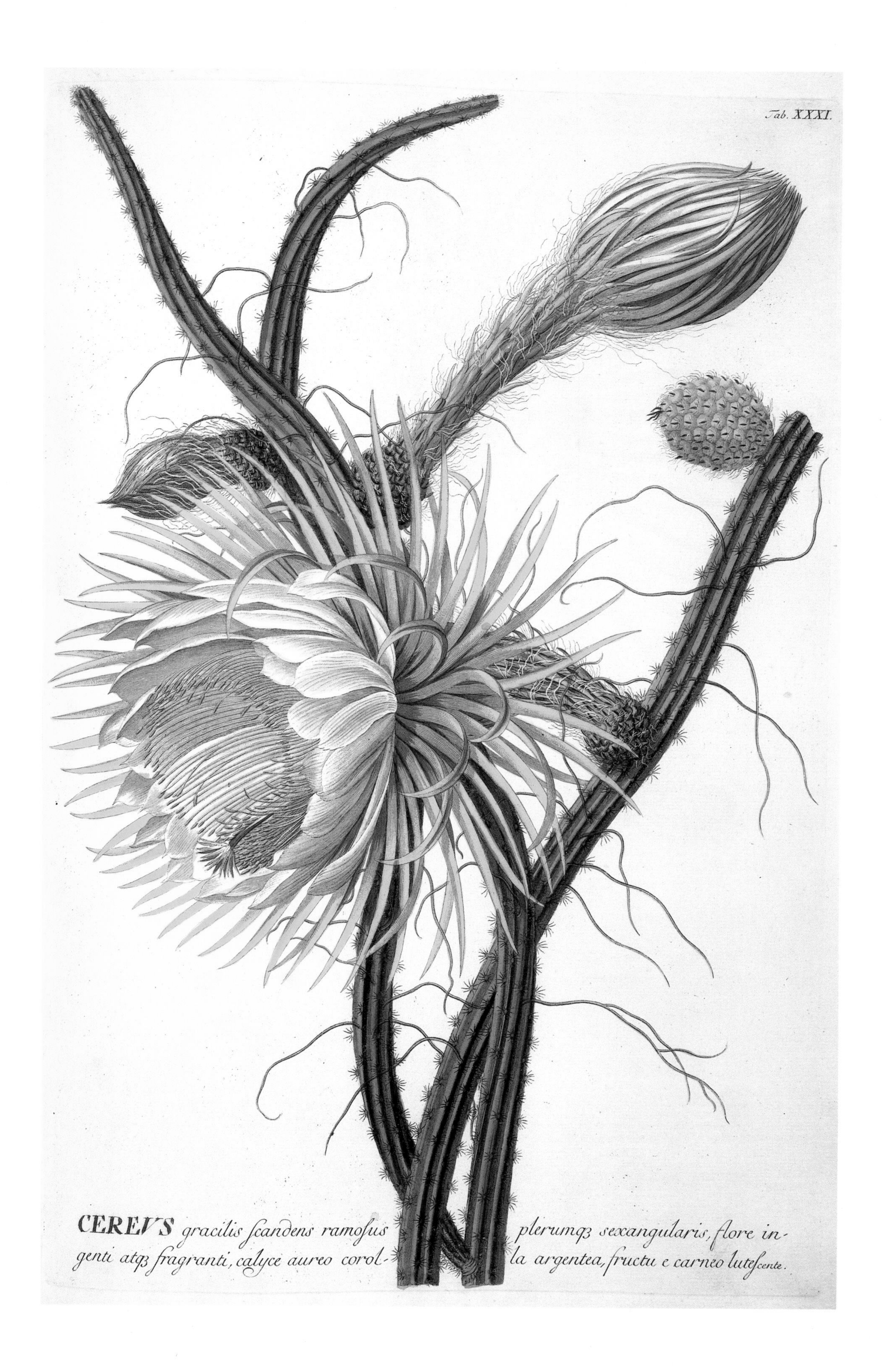

CEREVS *gracilis ſcandens ramoſus plerumqȝ sexangularis, flore ingenti atqȝ fragranti, calyce aureo corolla argentea, fructu e carneo luteſcente.*

impossible that they should be done from nature. They are fanciful arrangements, aided by drawings (of which not many survive), and finished during the long winter. This is an offence to the botanists, and has made it that their paintings have not yet been carefully and conscientiously examined, except by the Swedish Bergström, writing in that language. For the Dutch painters were friends of the great Dutch florists who sent them round their choicest flowers. Much botanizing, in point of history, remains to be done among the Dutch flower Masters. The Tulip of the Dutchmen could be redrawn and painted in all its fantasy and far-fetched colourings, a subject of exceptional fascination in itself;[3] and there are forgotten episodes of the carnation. In addition, there are the auriculas of the Flemings, the double poppies beloved of the Hollanders, and early stages of the ranunculus and hyacinth.

The course of the flower book is now set, evenly, between the latest discoveries from foreign lands and the florists' nurseries. Typical of this dual interest is Georg Dionysius Ehret (1708–70), the German flower painter, less exquisite than Nicolas Robert, as accurate as the early Dutchmen, but first and foremost a botanical draughtsman with the capacity to make drawings for great botanical works, and, therefore, to date, in many respects the most practical flower painter who has yet appeared. Whatever else his feelings or sensations Ehret was, certainly, not frightened of his flowers. He set himself to his task with appetite and gusto. But before discussion of Ehret, this is a convenient moment to mention a pair of other German botanical projects. They are the *Nürnbergische Hesperides* (1708) by Johann Volckamer, not a flower but a fruit book, uncoloured, and delightful for its architectural and garden backgrounds, some of them by Paulus Decker the Elder;[4] and the *Phytanthoza Iconographia* (1737–45) of the heavily bewigged Johann Weinmann, a rich pharmacist of Regensburg, legal capital of the Holy Roman Empire. Ehret was employed upon this. It is a hand-coloured work, enormous in scope and weight, and memorable after a dusty morning employed in turning over its millennium of plates for a section of Chinese-looking cactuses in china pots. Neither of these works, the *Nürnbergische Hesperides*, nor the *Phytanthoza*, is of greater interest than for its 'amusement content'. Ehret appears to have been scurvily treated over his drawings for the latter work, but soon received a roving commission from his friend and patron, Dr Trew of Nuremberg. This took him to Paris, to Holland, and to London, where he came to live in 1736. Here he remained for the second half of his lifetime, another thirty-five years, turning out his splendidly masculine drawings by the dozen, drawings which were 'big' enough to translate into engraving.

Plantae Selectae (1750–73) and *Hortus Nitidissimis* (1750–86) are Ehret's two masterpieces among flower books, and both are memorials of his close association with the learned Dr Trew. They were published long after Ehret had settled in England, and are to be regarded as projects slow in maturing and in execution which he had been preparing for many years. A preference may be expressed

PLATE 8
This cereus cactus was etched after a painting made by Georg Ehret for Christoph Trew's Plantae selectae *(Augsburg 1750–73). The inscription identifies the plant in a series of descriptive Latin phrases, a naming system simplified by Linnaeus with binomial nomenclature, or two-word species names. Ehret knew of Linnaeus's innovation, yet the artist persisted in the old way, with striking effect.*

for *Hortus*, since it contains so many of the florists' flowers, and of these Ehret was, it may be, the foremost painter that has ever lived. Not all of the plates in *Hortus*, however, are by him. But Ehret excelled in the tulip and the auricula. Nor did he despise the ranunculus and the polyanthus. And he painted these flowers before they had lost favour in the gardens of the rich. It would be impossible to exaggerate the beauty of Ehret's paintings of auriculas. In a Utopia of flower lovers, properly organized, there would be anthologies of his carnations, tulips, auriculas and ranunculuses, to go with similar collections drawn out of the imprisoned *florilegia*, and flanked by other forgotten beauties from a hundred sources, now neglected or ignored.

What is enjoyable in Ehret is the glorious health and sanity of his flower painting. In that world of the imagination wherein we arrange meetings of artists of the East and West how often have I wished that the florists' flowers of Northern Europe had fallen into Chinese and Japanese hands! And that Nicolas Robert or Ehret, in return, had drawn the flowering gardens of the Shogun at Kyoto, and of the Imperial Palace at Peking! In default of which it is of fascinating interest to study briefly the Chinese manuscript volumes of flower paintings, lately reunited by purchase, and now available in the Lindley Library at Vincent Square, Westminster.[5] For in comparison with Ehret and other draughtsmen, aesthetically, they engender another form of excitement altogether. They are paintings commissioned from Chinese artists early in the last century by the, then, Horticultural Society of London through the agency of an English merchant living in Canton. Among the paintings of tree peonies, flowers cultivated in Chinese gardens for many centuries, are some unknown or extinct varieties, a dark purple Moutan peony with a fringed edge like a pink, a double buttercup yellow of great clearness and brilliance, and a double kind with copper or cinnamon-coloured flowers. Nothing resembling these peonies is known to cultivation in our time. There are camellias drawn, curiously, as though they were hexagonal in shape; a pink-quilled chrysanthemum; a tree of chrysanthemums grafted in three colours in a pot, all branching from the one stem, or it could be that the artist has painted them, thus, merely for variety of display; and a drawing of quilled chrysanthemums that are like exploding rockets, like one of those late night rockets exploding high up after an interval of a second or two and coming down in rolled tubes or quills of coloured fire. Chinamen, it is well known, have a mania for fireworks, and the thought is irresistible that the painter had such displays in mind. We may wonder, indeed, what such painters would have made of the auricula which I would prefer painted, not by Chinamen, but by artists of Japan. It has been suggested already that Georg Dionysius Ehret was not frightened of his flowers but, neither, was he quite intoxicated by them. In his drawing he was sane and level-minded as though used and accustomed to them. When we remember that both the tulip and the ranunculus were brought into Northern Europe from Turkey it is not far-fetched to try and imagine that they could have fallen into the hands of the Oriental florists.[6]

With flower books, generally, there is the same difficulty that besets poets where it is a question of their Selected, or Collected Works. In the latter you are supposed to include everything, or nearly everything, that you have written and the poet is probably wiser who postpones this task and hopes someone else will take it on after he is dead.[7] But, in the case of Selected Works, they are usually prefaced with the old familiar formula that they include all that the poet 'would wish to preserve'. Where flower books are concerned, there are how many instances of every known flower of whatever genus being drawn and coloured in all their minuteness of differentiation, and instance after instance where selection would be preferable to entire collection! This is, of course, to look at such books from the point of view of aesthetics, not 'botanics'. For we are now approaching botanical works of 'exhaustive' range and execution, plate after plate of which can be turned over with indifference until we come to know that, as always, it is the artist who lives and who gives life to what is dead. Yet a good draughtsman can bring a dull subject alive, and in the field of botany it could even be said that there are in this respect more triumphs than there are failures.

There are minor flower painters of about the level of quality of the Chinese draughtsmen just mentioned who are able to sustain a wonderful temperature of interest throughout a monotonous subject. Such is Henry C. Andrews in his *Coloured Engravings of Heaths* (1794–1830) with its three hundred hand-coloured plates, and in his two large volumes on the South African Geraniums. But painters more considerable than Andrews devoted their talents to minute delineation. Such a one was James Sowerby (1757–1822), founder of a dynasty of painters. His scientific interests, which led him to make coloured drawings of minerals and crystals and which branched out in his sons and grandsons into works on conchology and ferns and grasses, all of prime quality and execution, are best studied in his own *Coloured Figures of English Fungi* (1797–1815), a borderline subject where this present work is concerned since funguses are not flowers. In more than four hundred hand-coloured plates it introduces one into the sub-human world of the *chanterelle* and toadstool, the scarlet *agaricus* and how many other wonders, edible or lethal? Black domes and towers of a Tartarean Kremlin; spotted phalli; and tattered funguses of mouldering flesh, still intact in rim or circumference, and recalling the skyline of Brueghel's *Triumph of Death*, in the Prado, where gaunt poles still uphold the rotting bodies broken on the wheel. This wonderful and grim work, in coloured inks, still fetches in the auction rooms but a fraction of what collectors are willing to pay for botanical works of far inferior merit.

Aylmer B. Lambert's *A Description of the Genus Pinus* (1803–24) is, in terms of aesthetics, another marvelous instance of controlled intoxication. We all know, that is to say, that no musician should ever play when he has had anything to drink. The Gypsy primas and the player of the cymbalom may be handed glasses of champagne to drink during the early hours of the morning but then the audience is under a spell and has lost its critical faculty. It is impossible to perform serious music in this manner. The virtuoso must eat and drink little, or nothing at all, before playing. Paderewski, it could be recalled, would come to the concert fasting, and before going on to the platform would plunge his hands into a bowl of nearly boiling water in the green room. Yet, in effect and nuance, he was the most subtly intoxicating of all pianists. But his hypnotic powers were the result of eight or even twelve hours practising a day, and of incessant attention to the slightest detail. It is, in part, by the reiteration of difficult passages until they become easy, or give the illusion of ease, that music becomes intoxicating and hypnotizing in performance. But close repetition is another of the narcotics that take effect in music; and most or all of these quasi-Oriental procedures in aesthetics are to be seen at work in

CORONA IMPERIALIS I.
William Rex
I.M. Seligmann excud. Norimb.
40.
I.C. Keller pinx. 1757

Lambert's *A Description of the Genus Pinus*. The plates produce their effect of stimulus and of slight intoxication by means of the incessant repetition of minute differences and variations in structure and in ornament. All but eight of the drawings are by Ferdinand Bauer; and of the aberrant number, one is by Ehret, and four by Francis Bauer, Ferdinand's elder brother. The *Genus Pinus* was the work so much admired by Goethe, but the brothers Bauer are but two more forgotten names among flower masters. Yet it is the question whether they were not the greatest of all botanical draughtsmen there have ever been. What is marvellous in the *Genus Pinus* is the play of the pine needles, wherein one soon comes to understand how it was that the Chinese *literati* evolved a whole school of artists who took for their subject nothing of more substance than the movement of bamboos.[8] It is, also, the incredible variety of tasselling in the branches; the diversity of the pine cones; even, the salubrity and pungency of the pinewoods with their therapeutic properties, as in the calm, clean air of the mountains in a hundred different lands, with the noise of the wind among the pine needles, but returning, always, to this incredible skill in their delineation where every individual needle leads its own independent life and has been given the importance of, it could be said in slight exaggeration, a sword blade by one of the great sword masters of the Far Orient. Ferdinand Bauer, and his brother Francis, are almost unique among painters in this capacity to invest a single plant form, or part of one, with prime importance.

Lindley's monograph on the Digitalis (1821) shows Ferdinand Bauer in just the same trauma of calm ecstasy before the foxglove. Each and every spire is drawn by him as though he were drawing Magdalen Tower or the Giralda of Seville. The marvellous stippling and mottling of each flower is rendered with patient accuracy, each mullion is drawn; and the spire of bells, when finished, could be an architect's despair but the pride and delight of the campanologist. Ferdinand was the traveller of the two brothers, accompanying John Sibthorp to the Levant to make drawings for the latter's *Flora Graeca* (1806–40), and going later with Matthew Flinders to Australia, a botanical project which petered out after a few engravings had been made, and Ferdinand returned to his native Vienna where some hundreds or even thousands of his unpublished drawings of Australian plants are still preserved. His brother Francis Bauer, perhaps the better artist of the two, at the instance of Sir Joseph Banks spent fifty years of his life as botanical draughtsman to Kew Gardens. Few, comparatively, of his drawings were engraved for books, but the heath *Erica sebana* from William Aiton's *Delineations of Exotick Plants* (1796) is a superb example of his powers (plate 18). Again the same value and importance is placed on every detail, as in his brother's *Genus Pinus*, so that each flower of the erica rides on the air as its own master. One or two more ericas in the same volume make of these humble heaths trees worthy to rank with the cedars of Lebanon. In their importance they are as centuries old cryptomerias, or the giant sequoias (redwoods) that, at midday, throw their tented shadows as high as themselves upon the California

PLATE 9
Christoph Trew illustrated Hortus nitidissimis *(Nuremburg 1750–86) with almost 200 prints made after his huge collection of botanical watercolors. "Corona Imperialis" was painted by J. C. Keller in 1757 and etched by Seligmann. The careful, almost opaque coloring has retained a startling freshness.*

IRIS Vulgaris Germanica, sive Sylvestris
Common German, or Wild Flower-de-luce
J. Edwards delin:
J. Fougeron Sculp

mountain sides, 'as high as themselves' meaning the height of, or all but as high as, the Pyramid of Cheops. More flaunting, by nature, are Francis Bauer's *Strelitzias*, a volume of lithographed plates brilliantly hand-coloured, ranking more, almost, as a bird book than a flower book, and making one wish that Ferdinand or Francis Bauer, either, or both of them, had been able to devote time and talent to all the varieties of hibiscus, flowers which float on the air, which ride the air as no other flowers do, and are now to be admired in their many hybrids in sub-tropical gardens among the flower inventions and improvements of recent years.[9]

Time now brings us to the threshold, or even upon the steps, of Dr Robert Thornton's *The Temple of Flora* (1799–1807), most famous of English flower books, the composite work of several painters, for Reinagle or Henderson were painters more than draughtsmen. Yet it is to be noticed that the less exotic flowers in *The Temple of Flora*, not the night-blowing Cereus, but the tulips, the hyacinths, the carnations, are, on the whole, by far the most successful of all the mezzotinted plates. The others are more than a little absurd in their extravagance taken, as they have to be, in company with Dr Thornton's verbiage, some passages of which, in their mixture of royal and tropical flavour, could as well be describing the chandeliers in Brighton Pavilion. The curious convention, surely an insistence on Dr Thornton's part, by which the flowers are seen at eye-level against a background of mountains, clouds, or moonlight, is reminiscent of an advertisement for a new scent, which described the sensation as comparable to that of 'waking up in a lily-field at dawn'. The most justly famous of the plates is that of tulips which, indeed, displays the flower in its forgotten faculties after two centuries of performance in the hands of Dutch, French, and English florists, and puts to shame the tame colours of the tulip as it is known to-day (plate 19). Two groups of auriculas are little less spectacular, while the glorious carnation plate has an additional interest because of the pair of blue 'flakes', a colour no longer found in carnations though its authenticity is vouched for in the scarlet and the crimson 'flakes' blowing near by. And yet . . . there is something in Dr Thornton's *The Temple of Flora* which could make one wish the paintings were by 'le douanier' Rousseau. Then, its exotic wonders would, indeed, have made a work of art. Dr Thornton who, it is hardly necessary to reiterate, ruined himself by his project, had one further contact with aesthetics when against his better instincts he employed William Blake to make a series of little woodcuts for a cheap reprint of his edition of Virgil for the use of schoolboys, little strips hardly longer than the adhesive tape on the back of an envelope, but pregnant with poetry and with enough 'message' to form a school of painters. Compared to *The Temple of Flora*, these works of art in their pastoral simplicity are like coming out of Brighton Pavilion into the humble cabin where Abraham Lincoln was born and spent his childhood in the wilds of Kentucky.

Equally sumptuous, Samuel Curtis's *The Beauties of Flora* (1806–20), with only ten plates, is so

PLATE 10

John Edwards produced several colorplate botanical books such as The British Herbal *(London 1769–70). This iris was etched in a minutely descriptive manner by J. Fougeron after a drawing by the author. It is colored with equal care, and bears comparison with our plates of other contemporary works (Plates 13–15).*

excessively rare that it is hardly known at all to the public. Many wonderful things are hidden here, most of them painted by Clara Maria Pope. Of all little known English flower books this is perhaps the most splendid. Tulips in no way inferior to those of Dr Thornton; pinks and carnations; glorious auriculas and polyanthuses; a plate of double hyacinths, almost the sole remaining relic of the hyacinth rage which during the eighteenth century was nearly on a par with the tulipomania of a hundred years before, and exhibiting such hyacinths as no living eyes have seen; staid ancestors of the dahlia; a dazzling and intoxicating array of the ranunculus in all its forgotten glory;[10] and a plate of anemones in a multiplicity of forms and colours, the results of continual experiments over several human generations in France and Holland, something else lost to the modern world and revealing new beauties of colouring and structure in the anemone with no sacrifice of its primal grace and innocence.[11]

It seems appropriate to mention next the *Pomona Britannica* (1812) of George Brookshaw, for it is a work on the same scale. There are peaches and plums to make the mouth water, melons and greengages, all set down carefully in classical drawing as though the artist was copying the frescoes of Pompeii. Another and smaller fruit book, Thomas Knight's *Pomona Herefordiensis* (1811) is a cyder feast of all the local apples, described under their poetical old names, and beautifully drawn and coloured mostly by a lady amateur. To her company, and that of the talented Clara Maria Pope, must now be added the person of a lady painter of Liverpool, Mrs Edward Bury, amateur artist of *A Selection of Hexandrian Plants* (1831–34), hippeastrums, crinums, pancratiums, the best of them being her several plates of crimson 'lillies', all engraved for her by Robert Havell, one of the masters of the English school of aquatint. This is a valuable book, but amaryllises apart, repetitive and monotonous.

In France it was the age of Pierre Joseph Redouté (1759–1840); and he is now so much the most famous of the flower masters because of his roses, that it is a delight to have in this present volume three plates from his *Les Liliacées* (1802–16) as well as two of the most beautiful of all his rose drawings. It is difficult to add more to what has been said about Redouté already. His books are masterpieces among flower books, but from examination of his original drawings he does not emerge as a greater flower painter than two or three other artists contemporary to him who are almost forgotten. Redouté must have had personality, in life, and he got his works published. Yet, he was both a tireless worker and hopeless over money. But it all adds up to his being Redouté, while Gerrit van Spaëndonck remained van Spaëndonck, and Turpin, Turpin. It is significant that the splendid *Fleurs dessinées d'aprés Nature* (1801) is the only work of van Spaëndonck published in his lifetime. He is another of the painters who is to be studied in his *Vélins* at the Jardin des Plantes. Jean-Louis Prévost's *Collection des Fleurs et des Fruits* (1805), like van Spaëndonck, is another work of the Napoleonic age which is alone in its eminence, being at least equal to Redouté. As to Turpin, he is to be admired in Duhamel du Monceau's *Traité des Arbres Fruitiers* (1768), also in works figuring American plants collected by Humboldt and Bonpland; but, somehow, Turpin never managed to attract as much attention as Redouté. His collaborator in the book on fruit trees, Pierre-Antoine Poiteau, illustrated a work as fine and compendious as any of those named, J. Antoine Risso's *Histoire Naturelle des Orangers* (1818), which could only fairly be described as a Hesperidean holiday among the orange and citrus groves, for it contains exquisite drawings of every known variety of orange, lemon and grape fruit, and their congeners, fruits that hang from the leaves, alternately, like suns or moons, with every kind of rind, and shaped like gourds or pitchers, misshapen and peculiar, or again, authentic globes of fire, whether pale, as of moonlight, or red-gold like the sun but half-hidden, as in poetry, in its own green shade. A beautiful and inspiring work, in its way not less so than Redouté's *Les Liliacées* (1802–16) or *Les Roses* (1817–24).

If we return, now, to English flower books, for time is running low, it is to be confronted with James Bateman's *The Orchidaceae of Mexico and Guatemala* (1837–43), an elephant folio as large and heavy as Audubon's *Birds of North America*, their only rival for sheer size being that work on the coronation of the Tsar Alexander II which is so enormous that no one man can lift it, and it has to be wheeled to your desk in the North Library of the British Museum Reading Room upon a trolley. Surprisingly, the *Orchidaceae* is a work of high quality. The hand-coloured lithographs are after drawings by two lady painters, and it is delightful, in this Gargantuan project, to find the hand of George Cruikshank in it, for Cruikshank has the 'midget' touch and is the most Lilliputian of the old masters.

A botanical draughtsman now appears, who, it could be said, would be somewhat of a parallel to Gould among the bird books of the Victorian age were it the case that Gould had made his own illustrations for his works. Gould, though, did not draw his own plates. A situation, in short, which could be adapted to a well-known phrase of Groucho Marx, and would then run, 'Everything about Fitch reminds me of Gould, excepting Gould'. For this draughtsman is Walter Hood Fitch (1817–92). He shows his hand, as did Gould, in everything and there is hardly a botanical project during half-a-century of Queen Victoria's reign in which he was not involved. For forty years he made drawings for *The Botanical Magazine*, under the editorship of Samuel Curtis and others, but his finest achievements were in association with Sir Joseph Hooker for whom he redrew and lithographed the splendid series of plates for *The Rhododendrons of Sikkim-Himalaya* (1849–51), and *Illustrations of Himalayan Plants* (1855). As well, he made the illustrations for Hooker's *Victoria Regia* (1851), a monograph on the great waterlily from the Amazon, a sort of botanical *Great Eastern* and one of the wonders of the age. Fitch had the greatest competence of any botanical painter who has yet appeared in drawing the rhododendron. It will be seen from his illustration that he isolates it and draws it calmly as though it is a chestnut-candle. His plate of *Magnolia Campbellii* does full justice to this wonderful rose-pink Himalayan. And finally, for like Gould he is ubiquitous, Fitch drew and made the lithographs for Elwes's *A Monograph of the Genus Lilium* (1877–80), a work that could be called a pendant to Bowdler Sharpe's two volumes on the birds-of-paradise and is one of the culminating ornaments of the Victorian age. It is this work which has been brought up to date in our own times by Grove and Cotton's supplement illustrated by Miss Lilian Snelling.

A word must be said concerning the florists' magazines. It is among these that we look for the tulips, auriculas, carnations of the second florists' age. They are a mine of curious information wherein, for instance, we may learn how tulips were divided into bybloemens, prime baguets, rigauts, incomparable verports, roses and bizarres; bybloemens, in detail, having white bottoms marked with various colours, and bizarres having yellow grounds. I have a notebook filled with notes on *The Florists'*

Magazine (1825), *Floricultural Magazine* (1837), *The Floral Cabinet*, *Floricultural Cabinet*, and *The Florist*, even penetrating to so remote a fastness as *The Gooseberry Growers' Register* (1839–43), giant gooseberries of the Cheshire and Lancashire fanciers, bigger than golf balls, and growing on bushes that resembled dwarved Chinese trees. There are the Chellaston Bybloemens for which Derbyshire was famous, the *Bouquet Pourpre* hyacinth with wedge-shaped dashes of brilliant green in it, and how many other wonders! A tulip called *Rose Camuz de Craiz*, scarlet and white; among carnations of the show bench, that is to say, with trimmed petals, and mounted on paper, the tremendous *Jenny Lind* and *Justice Swallow*, *The Emperor of China*, *Roi des Capuchins*, and *Colonel of the Blues*; and among auriculas, *Taylor's Glory*, and *Page's Waterloo*. For another, and little marvel there could be the ranunculus *Œillet Parfait* from James Hogg's *The Florists' Guide* of 1827–29. There are, as well, beautiful flowers of these named sorts in Mrs Loudon's *Ornamental Perennials* (1843), and in the six volumes of Sweet's *The British Flower Garden* of 1829–32. But the appeal of these flower plates is in the nature of a *harlequinade* or *charivari*, and it is as such that they should be assembled in their companies and platoons. There, they are of the same band of players as J. M. Kändler's Meissen comedians, or Franz Anton Bustelli's troupe from Nymphenburg, that is to say, lovely and immortal in their chequered suits. And they end the long trajectory with their brilliant posturings and somersaults.

There is, happily, no reason to think that the fine flower book is ended. For additions are made to it as the years go by. There are flower painters at work, now, who are little, if at all, inferior to their forbears. Photography, except in one or two particular instances where the flower and background are suited to the medium, as in a recent work on cactuses, can never possibly equal the drawn and coloured portrayal by the human hand. We have only to consider what would be the dull and blurred uniformity of coloured photographs of, let us say, irises, compared to the magnificent effect they could give in drawing. It has been suggested, already, that fritillaries would furnish one of the most beautiful of all themes for a flower painter. And in this connection it is not altogether inappropriate to recall the much derided remark of Liszt, in his *Life of Chopin*, that he would like to harness a considerable pianist to every individual one of Chopin's Mazurkas. Their incredible rhythmical differences and depths of nuance deserved, he thought, so extreme a degree of specialization. Even so with the fritillary which is, at once, a momentary vision and the study of a lifetime. In the same spirit we look at a drawing of a flower by Leonardo, by Pisanello, by Albrecht Dürer, and turn to the great flower books. They are but likenesses, or counterfeits. But they do not drop their flowers, or lose their freshness, and the pleasure lasts for a short life or for a long one.

PLATE II

John Miller classified the plants at Chelsea Physic Garden in London according to Linnaeus's system based on flower structures. Illustratio systematis sexualis linnaei *(London 1770–77) includes Miller's marvelous etching of this New World species, "Helianthus," the sunflower, to illustrate "Class 19, Order 3" of Linnaeus's system.*

Leontodon Taraxacum.

THE ILLUSTRATORS OF THE GREAT FLOWER BOOKS

Wilfrid Blunt

THE SPLENDID AND SUMPTUOUS FLOWER-BOOKS now so prized by collectors are for the most part the product of the eighteenth and nineteenth centuries. It must not, however, be forgotten that a printed herbal illustrated with figures of plants appeared as early as 1481, and we must therefore pass in review some of the achievements of an earlier age.

Incunabula herbals were in general illustrated with stylized figures ultimately derived, after generations of copying and recopying, from drawings made in a remote past from living plants. Very occasionally, an artist might find himself obliged, for lack of a figure to copy, to draw a plant direct from nature; a few crude though honest woodcuts of this kind are to be found among the pages of the *German Herbarius* of 1485. It was not, however, until the publication, in Strasbourg in 1530, of the first volume of Brunfels' *Herbarium Vivae Eicones* that this deplorable tradition of copying was seriously challenged. The title of the book—'Living portraits of plants'—explains the aim of its illustrator, Hans Weiditz. Throughout, the flowers have been carefully drawn from nature. It marks the beginning of a new era in the printed botanical book.

Far more impressive in appearance, though less sensitive in execution, is the great folio herbal, *De Historia Stirpium*, of Leonhart Fuchs, published in Basle in 1542. With its hundreds of full-page illustrations of plants, it deservedly ranks as the first of that long line of monumental flower-books which during the last four hundred years have poured from the printing-presses of Europe. Many other fine herbals—those, for instance, of the Italian physician Pierandrea Mattioli—were issued during the sixteenth century, but none was quite so lavishly illustrated.

Towards the close of the century, two important factors affected the development of botanical illustration. First, the substitution of the metal plate for the woodblock made possible a new delicacy of line; and second, the growing interest in garden flowers, as opposed to merely useful plants, provided artists with a new class of patron and more spectacular material to portray. One of the loveliest of these 'florilegia'[12] was the *Hortus Floridus* (1614) of the Dutchman Crispijn vande Pas; a more sensational production was the gigantic *Hortus Eystettensis* (1613) of the Nuremberg apothecary Basil Besler. Nor should we overlook the part played by France, where Pierre Vallet was responsible for the delightful *Le Jardin du Roy très Chrestien Henri IV* (1608) and Paul Reneaulme (or the artist who worked for him) for the delicately etched illustrations to his *Specimen Historiae Plantarum* (1611).

All these books are, alas! outside our terms of reference. So, strictly speaking, is Nicolas

PLATE 12

William Curtis's Flora londinensis *(London 1775–87) has 432 prints of wildflowers growing near London. A folio sheet allows for the life-size representation of "Leontodon Taraxacum," a dandelion painted and etched by William Kilburn. The plates were offered plain, at half the cost of colored plates, and sets with exceptionally fine coloring cost three times that of the uncolored plates.*

Robert's *Recueil des Plantes*, the plates for which were made during the last three decades of the seventeenth century. However, since all but about forty of the illustrations remained unpublished until 1701, it is perhaps permissible to stretch a point and consider the work as included in our period. With these three noble volumes, therefore, we may begin a more detailed survey of the fine flower books produced between 1700 and 1900.

In most countries where the florilegium flourished in the seventeenth century, private patronage paid the piper and called the tune. In France, however, official Court Flower Painters were appointed, with regular salaries and under the obligation to produce a stipulated number of paintings each year. One of these was Nicolas Robert (1614–85), the son of a French innkeeper at Langres. Robert had made his reputation as the illustrator of the famous *Guirlande de Julie*—a little anthology of flower-paintings and poems presented by the baron de Sainte-Maure to his fiancée, Julie d'Angennes, before he set out for the wars. This pretty book had brought him to the notice of Gaston d'Orléans, Louis XIV's uncle, who was seeking a painter to record the rare flowers in his garden at Blois. Shortly after Gaston's death in 1660, Robert was appointed 'peintre ordinaire de Sa Majesté pour la miniature'. When the newly-founded *Académie Royale des Sciences* decided to publish an illustrated History of Plants, Nicolas Robert was very naturally chosen as principal artist, his collaborators being Abraham Bosse and Louis de Châtillon. In all, 319 plates were engraved, thirty-nine of them being used to illustrate Denis Dodart's *Mémoires pour servir à l'Histoire des Plantes*, which was published by the Royal Press in 1675. But before the whole set was ready, war had put a brake on such lavish productions. In 1701, the plates were issued without letterpress, and it was not until about a century after its inception that the book was finally published in the form originally intended. Meanwhile botanical details, very mechanically engraved, had been added to the plates. Since the volumes were never sold to the public, they are now a considerable rarity; it may, however, be worth mentioning that the copper plates were not destroyed, and modern pulls from them, almost indistinguishable from the original prints, have been made.

Dodart has clearly defined, in the preface of his *Mémoires*, the task set to his artists. Great importance was attached to the plants being portrayed their natural size. Those rather too large for the page were represented as cut in two, those yet larger being shown with some detail full-scale. Colour-printing being not yet available, and hand-colouring being considered too laborious and too unreliable, every effort was made to indicate at least the depth of the colour of flowers and leaves. Robert undertook the lion's share of the engraving. Châtillon was particularly successful with small, delicate plants, Bosse with the coarser and more robust. As for Robert, he could be relied upon to make a fine engraving of any flower that was put before him. The *Recueil des Plantes* is an important landmark in botanical

PLATE 13

Pierre Buc'hoz produced folio-sized books which only sought to entertain.
Collection précieuse *(Paris c. 1776) includes this hand-colored etching of a peony. Text is limited to an index that often records the native countries of these flowers.*
Each plate has a border echoing the gold leaf found on natural-history watercolors kept by royal horticulturists.

illustration and a noble example of book production. Van Spaëndonck, writing at a time when comparison could be made with the works of such great artists as Ehret and Redouté, still considered it the finest flower-book that he knew.

It was very natural that the Dutch, who throughout the seventeenth century showed a passionate, at times almost hysterical, enthusiasm for horticulture, should have produced a large corpus of flower-paintings and flower-books. From among the latter we have selected for illustration—though it is in the first instance concerned with insects—Maria Sibylla Merian's *Metamorphosis Insectorum Surinamensium*.

Maria Sibylla Merian (1647–1717), the daughter of a Swiss engraver, was born and educated in Germany; her mother, however, was Dutch, and on the death of Matthaeus Merian his widow married the Dutch flower-painter Jacob Marrell. Maria herself became the wife of a German artist, and her first insect and flower books were published in Germany. But in 1685 she became a convert to Labadism,[13] left her husband, and settled in Holland. Three years later, accompanied by her younger daughter Dorothea, she sailed for Surinam (Guiana) where she collected the material for her greatest work, *Metamorphosis Insectorum Surinamensium*, which was published in Amsterdam in 1705. Though Maria Merian's first concern is with the insects that she portrays, the plates also show the plants upon which these insects live; these plants are drawn with the same delicacy and precision as the insects themselves, and the book may thus legitimately be considered a florilegium also. She has wonderfully conveyed the exotic colours of these tropical flowers, moths and butterflies, at that time almost wholly unknown to Europeans. Maria Merian was a fine artist. But, effective as the hand-coloured plates of the *Metamorphosis* are, they give a faint idea of the meticulous craftsmanship that is to be found in the original drawings. These, together with replicas of them probably made by her daughter, are in the Royal Library at Windsor and in the British Museum, and more of her work is at Rosenborg Palace in Copenhagen, and in Leningrad.

No mention has yet been made of English flower-books. The two most popular early works were undoubtedly Gerard's *Herball*, first published in 1597 and reissued, with revised text and new plates, by Johnson in 1633; and Parkinson's *Paradisus* (1629). These were illustrated from wood-blocks borrowed from the Continent or adapted from foreign designs. It was not until the eighteenth century that England made any important original contribution to botanical illustration, and even then her first considerable artists were foreigners who settled in this country. Among these was Jacob van Huysum, a younger brother of the famous Dutch flower-painter, who made most of the drawings for John Martyns' *Historia Plantarum Rariorum* (1728–36) and the *Catalogus Plantarum* published by the Society of Gardeners in 1730. Their particular interest lies in their being illustrated by an elementary kind of mezzotint printed in colour; thus they may be considered the humble ancestors of such splendid productions as Thornton's *Temple of Flora*.

As a representative English book of the first half of the eighteenth century we have chosen Robert Furber's *Twelve Months of Flowers*, first published in 1730 and subsequently reissued in several smaller and relatively worthless editions (plate 5). Robert Furber was a nurseryman, and his book nothing more nor less than a sumptuous catalogue of the plants that he stocked. The flowers are grouped in the manner of a Dutch flower-piece, each urn containing some thirty different kinds which (allegedly) bloomed in a particular month of the year. The original paintings were the work of the Flemish artist Pieter Casteels; they were engraved by H. Fletcher and coloured by hand. The book is of importance as a record of the plants cultivated at the time. First in popularity comes the auricula, with twenty-six named varieties, followed closely by the anemone, hyacinth and rose; twenty-five American plants are included, among them being the Virginia aster (*Aster grandiflorus*). But considered as works of art, the plates, which inevitably invite comparison with Dutch flower-pieces, are disappointing, and the high price that the book now commands is out of all proportion to its merits.

A far more considerable foreign artist who made England his home was the German-born Georg Dionysius Ehret (1708–70), the son of a Heidelberg market-gardener. Work as a garden boy with an uncle at Bessungen proved a ceaseless grind which almost broke his spirit and afforded him little leisure to pursue his hobby of flower drawing, but promotion to the charge of one of the gardens of the Elector of Heidelberg gave him his opportunity; here he attracted the attention of the Margrave of Baden, who carried him off and set him to paint the flowers growing in his own garden. After various vicissitudes, the young man made the acquaintance of a wealthy Nuremberg physician, Dr Trew, who was to become his lifelong friend and most influential patron. To his liberality we owe the two fine florilegia, *Plantae Selectae* (1750–73) and *Hortus . . . amoenissimorum Florum imagines* (1750–86); a page from the former is illustrated in this book (plate 8). Passing by way of Switzerland, Ehret reached Paris where he was kindly received by Bernard de Jussieu who urged him to reconsider his intention of going to Holland and to seek his fortune in England. After visiting England, and Holland where he met Linnaeus and the Dutch banker Clifford, Ehret returned to England and settled there permanently. His success was immediately assured: the Duchess of Portland, Sir Hans Sloane and Dr Mead recognized his talents and were active in furthering his interests. Besides his more serious work, he gave lessons in flower-drawing to the daughters of the English aristocracy in London during the season, and in the summer made a round of their country houses. In 1757 he was elected a Fellow of the Royal Society.

Ehret's best work is his series of gouache drawings of flowers, magnificent examples of which can be seen in the Natural History and Victoria and Albert Museums, London, and the Kew Herbarium. Many of these were engraved to serve as illustrations to his books, of which the finest are his own *Plantae et Papiliones Rariores* (1748–59) and the two volumes published by Trew. Botanically interesting though aesthetically inferior are the illustrations that he made for various books of travel such as Pococke's *Description of the East*. *Plantae et Papiliones Rariores* is a slim folio containing fifteen plates engraved by Ehret himself and very delicately coloured by hand. A more robust volume is Trew's *Plantae Selectae*, with its fine title-pages and mezzotint portraits of Trew, Ehret, and J. J. Haid who with his son was responsible for the engraving of the plates. That of *Lilium superbum*, a magnificently decorative sheet, was made from a drawing of a specimen flowering in August 1738 in Dr Collinson's

garden. Not all, however, of the hundred plants figured in Trew's book will excite the lover of fine flower-prints, for a number of them have little to recommend them beyond their novelty. More uniformly interesting to both artist and gardener are those shown in the three folio volumes of Trew's *Hortus Nitidissimis*, a book which aimed at presenting 'a complete collection of the most magnificent tulips and crown imperials; the sweetest hyacinths, daffodils, narcissi and jonquils; the most charming roses, carnations and snowflakes; and the loveliest lilies, fritillaries, ranunculuses, anemones and auriculas'. They constitute a valuable florilegium of the plants, especially the florists' plants, of the gardens of that time. Other German artists besides Ehret contributed drawings, that of the Hyacinth "Gloria Mundi" being the work of F. M. Seligmann of Nuremberg.

Ehret's greatest merit is that he succeeded, as few other botanical artists have succeeded, in being at once both botanist and artist. His work is splendidly vigorous, yet he does not sacrifice botanical accuracy. His engravings and paintings make wonderful wall decorations, but at the same time they reveal an intimate knowledge of plant structure and habit. His preference for gouache gives his original paintings a strength that cannot be achieved by the transparent washes of a Redouté; and many, though not all, of his engraved plates are also retouched with opaque colour.

Another German artist who made England his home was Johann Sebastian Müller of Nuremberg—or John Miller, to use the name by which he was generally known in this country. He was responsible for the fine plates in *An Illustration of the Sexual System of Linnaeus* (1777), of which we reproduce the giant sunflower (plate 11). Linnaeus himself considered them 'more beautiful and more accurate than any that had been since the world began'; though unequal in quality, the best of them are certainly impressive. More than a thousand of Miller's original drawings are in the Natural History Museum, London, and the Lindley Library of the Royal Horticultural Society contains a copy of his rare *Botanical Tables*, financed by Lord Bute and published at vast expense in an edition of only twelve copies.

In France, the tradition of fine flower-painting established in the seventeenth century had been maintained by a succession of official Court painters, of whom Claude Aubriet was probably the most brilliant; his published work however, exquisite though it is, lacks the spectacular quality of his original paintings and could scarcely claim a place in a volume dedicated to Great Flower Books. During the second half of the eighteenth century were published François Regnault's *La Botanique* (1774) in three folio volumes, de Sève's *Recueil de vingtquatre Plantes et Fleurs* (c. 1772), Pierre Bulliard's *Herbier de la France* (12 vols., 1780–95) and Gautier-Dagoty's *Collection des Plantes* (1767) with remarkable plates printed by the Le Blond three-colour process. All these are important books, yet none of them can stand comparison with the *Fleurs dessinées d'après Nature* of Gerard van Spaëndonck.

PLATE 14

The illustrations etched for Pierre Buc'hoz's Collection précieuse *(Paris c. 1776) rely on bright hand-coloring for their beauty. For this white amaryllis, the author provides a blue wash background to help distinguish it from the paper.*

N.o 13
John Edwards delin.t et sculp.t
Large Orange Lilly
Publish'd as the Act directs June 1 1783

Gerard van Spaëndonck (1746–1822), by birth a Dutchman, established himself as a painter of flower-pieces in the manner of Jan van Huysum before he set out to seek his fortune in Paris. At the age of thirty-four he was appointed to succeed Madeleine Basseporte as *Professeur de peinture de fleurs* at the Muséum of the Jardin des Plantes. Perhaps his success as a teacher, his skill in producing elegantly decorated little snuff-boxes and other trifles for the ladies of the Court, and the labour of painting his annual quota of botanical studies for the official collection of *vélins*, account for the fact that his published work (except for a very inferior posthumous volume) is limited to a single book: the *Fleurs dessinées d'après Nature*, which appeared about the year 1800. This wonderful folio contains twenty-four magnificent stipple-engraved plates, the engraving being chiefly the work of P. F. Le Grand. Copies are occasionally found either hand-coloured or printed in colour and retouched with water-colour; though collectors not unnaturally pounce upon these rarities, the water-colour serves in fact to cloak the delicacy of the stipple-work. In their black-and-white state, these prints may well claim to be the finest flower-engravings ever made.

Van Spaëndonck's name is little known to the general public; that of his pupil, Pierre Joseph Redouté (1759–1840), though his talents were no greater, has become a household word. Possibly he was more industrious, probably he was more ambitious, and certainly he had the good fortune to attract the attention of the most influential patronesses in France. Redouté was a Belgian by birth (or perhaps one should say a Luxembourger, for the town of St. Hubert, where he was born, was at that time still within the borders of the Grand Duchy). Like van Spaëndonck, he too fell under the spell of van Huysum and also came as a young man to seek his fortune in Paris, where his elder brother was already employed as scene designer to the Théâtre Italien. At the Jardin des Plantes Pierre Joseph made the acquaintance of the botanist L'Héritier de Brutelle, and it was probably he who obtained for the young artist the position of Draughtsman to the Cabinet of Marie-Antoinette. No less important, however, was the influence of van Spaëndonck, who developed his technical skill while L'Héritier enlarged his botanical knowledge.

Redouté's great chance came when, in 1798, Josephine Bonaparte acquired Malmaison and resolved to establish there a garden for rare plants. As was the fashion, she searched for an artist to make records of them, and her choice fell upon Redouté. Thus were born Ventenat's *Jardin de Malmaison* (1803–04) and Bonpland's *Description des Plantes rares cultivées à Malmaison et à Navarre* (1812–17), both superbly illustrated in colour by Redouté. To the same period belong the eight magnificent volumes of *Les Liliacées* (1802–16), which were followed by the still more famous *Les Roses* (1817–24). Two of the plates reproduced in this volume are taken from *Les Roses*.

PLATE 15
John Edwards illustrated A Collection of Flowers *(London 1783–95) with partly color-printed etchings of his own making. Edwards hand-colored the seventy-nine prints as well, achieving a rare delicacy. "Orange Lily" is a stunning example of his work.*

The story of Redouté's life, and the unflattering description of his personal appearance, have been recounted too frequently to bear repetition here. During the First Empire, he won fame and wealth. Distinguished pupils flocked to his studio for tuition. But he was incapable of living within his means, and towards the end of his life found himself on the verge of bankruptcy. On June 19th, 1840, while he was examining a white lily brought to him by a pupil, he had a stroke. The following day he died. Upon a wreath of lilies and roses laid on his coffin were inscribed the following words:

O peintre aimé de Flore et du riant empire,
Tu nous quittes le jour où le printemps expire.

As we have already said, Redouté was singularly fortunate in the patronage he enjoyed; he was no less lucky in having the assistance of a brilliant team of stipple-engravers to transfer his drawings to the copper plate. Curiously enough, stipple-engraving, which in England was so successfully used for portraiture, was rarely employed there for botanical illustration, to which it was so eminently suited. Redouté came into contact with the process when he was in England in 1786, and himself perfected a method which his assistants employed under his direction and which earned him the award of a medal from Louis XVIII.

Of the many other fine French flower-books of the first half of the nineteenth century we have selected for reproduction a page from Jean-Louis Prévost's *Collection des Fleurs et des Fruits* (1805; plate 28). The book was compiled to assist designers of china, toiles and chintzes and is not primarily botanical in intention. The most striking plates show several different flowers grouped together in a bouquet; but though the arrangements may appear artificial, and the flowers doubtfully matched as to season, the drawing is accurate and the observation exact. With the death of Redouté in 1840, the age of the spectacular French flower-book drew to its close. Many excellent and highly-skilled botanical draughtsmen continued to produce important scientific books; but these are not of the kind to excite the cupidity of the collector who prizes the work of Prévost and Redouté.

From France we turn once again to England where, towards the close of the eighteenth century, two further continental botanical draughtsmen established themselves and set a standard of excellence that our finest native flower-painters were unable to rival. These were Francis (Franz) Bauer (1758–1840) and his brother Ferdinand (1760–1826), of Feldsberg near Vienna.

In Vienna, Nikolaus von Jacquin—a Dutchman of French origin—had gained the position of leading botanist of his day. Like Trew in Nuremberg, or Sloane and Banks in London, Jacquin was the centre of a flourishing circle of scientists and artists to whom he extended his patronage, and many handsome books owed their existence to his enthusiasm and enterprise. When John Sibthorp, Sherardian Professor at Oxford, arrived in Vienna in 1784 on his way to the Levant, Jacquin introduced him to Ferdinand Bauer who agreed to accompany the English botanist to Greece. The fruits of this journey were the ten celebrated volumes of the *Flora Graeca* (1806–40), which were produced, after Sibthorp's death in 1796, by Sir J. E. Smith and John Lindley. Only twenty-eight copies of the first edition were issued, at £254 a set, and the cost of the undertaking, estimated at about £30,000, was in the main defrayed by money bequeathed by Sibthorp. Nearly a thousand original drawings made for the *Flora Graeca*, together with three unpublished volumes of drawings of Greek fauna and a series of landscape sketches, are in the Oxford Botanic Garden library. Some of these landscapes were used for the title-pages.

Ferdinand Bauer, who after his return from the Near East had made England his home, did not accompany Sibthorp on his second trip to the Levant. But the wanderlust was in his blood, and in 1800 he sailed with Matthew Flinders on his adventurous expedition to Australia, from which he returned five years later with a large collection of drawings. Some of these he himself engraved for his *Illustrations Florae Novae Hollandiae* (1813), a work which he was forced to discontinue after fifteen plates had been published. Discouraged by his failure, which must be attributed in part to the Napoleonic Wars and in part to his own high standards which prevented him from employing the assistance of other engravers, he returned to Austria where thirteen years later he died. Among other books to which Ferdinand Bauer contributed drawings were the *Historia naturalis Palmarum* (1823–50) of the Munich botanist Karl von Martius, and the *Digitalium Monographia* (1821) of John Lindley.

We have dealt first with the career of the younger of these brilliant brothers. Tame by comparison was that of Francis who, attracted to England no doubt by Ferdinand, settled at Kew where he remained for fifty years making patient records of the plants that adventurous travellers sent back from all the quarters of the globe. One of these was Francis Masson, the first collector sent out from Kew, whose heaths from the Cape were painted by Francis Bauer for his *Delineations of exotick Plants cultivated in the Royal Garden at Kew*, published by W. T. Aiton in 1796. Among other works illustrated by Francis is his *Strelitzia depicta* (1818), with elaborately hand-coloured lithographs of all the known species of the genus. Francis died at Kew in 1840 and was buried in the churchyard there, by the side of Gainsborough and Zoffany.

Francis and Ferdinand Bauer may well claim to be the greatest of all botanical draughtsmen. Their skill in the execution of detail is miraculous, yet they never lost sight of the wood for the trees; everything is understood, balanced, controlled. To appreciate their true genius, it is necessary to examine their original drawings, some hundreds of which are stored in the botanical library of the Natural History Museum. In their engraved work, Ferdinand fared in general more happily than Francis. The splendid illustrations to Lambert's *A Description of the Genus Pinus*, all but eight of which were his work, deeply impressed Goethe. 'It is a real joy', he wrote, 'to look at these plates, for Nature is revealed, Art concealed'. The botanical draughtsman was no longer the mere recorder of floral beauty; he now had the more difficult task of serving both Art and Science. This, thought Goethe, might well seem an impossible undertaking. The drawings of Ehret, and more particularly those of the two Bauers, are proof that it is not.

Our consideration of English botanical books illustrated by artists of foreign birth may have given the false impression that in the eighteenth century English native talent was wholly non-existent. It is, however, true to say that most of the plates illustrating works such, for example, as Mrs Blackwell's *A*

Various Tulips.

Curious Herbal (1737–39) or Catesby's *A Natural History of Carolina* cannot compete with those made by continental-born draughtsmen. But under the direction of William Curtis there came into being a school of skilful English botanical artists, and since 1780 there has been no lack of local talent.

William Curtis (1746–99) was trained as an apothecary. But his heart was in botany, and at the age of twenty-seven he was sufficiently highly thought of in this field to be appointed Praefectus Horti and Demonstrator to the Society of Apothecaries at Chelsea. His principal interest was the British flora, and with the support of Lord Bute he soon after began to accumulate material for a series of folio volumes describing and illustrating the wild flowers growing within ten miles of London. The first part of the *Flora Londinensis*, as it was called, was published in 1777. Realizing the magnitude of his undertaking, Curtis resigned his post at Chelsea and devoted himself for the next ten years to his great work. By then, seventy parts had been issued, with a total of 434 plates; but financially the venture was proving disastrous, and Curtis was forced to discontinue publication. Fortunately for him he was clever enough to see where the trouble lay. Gardeners wanted pictures of garden flowers and had little interest in mere 'weeds'; they, rather than the botanists, could afford to pay, and he must humour their fancy. In 1787, he began the publication of *The Botanical Magazine: or Flower-Garden Displayed*, a periodical designed to illustrate and describe 'the most Ornamental FOREIGN PLANTS, cultivated in the Open Ground, the Green-House, and Stove'. It was the first, and has remained the greatest, of numerous botanical and horticultural periodicals, and with barely an interruption has continued publication down to the present day. As Curtis himself said, the *Botanical Magazine* brought him 'pudding' whereas the *Flora Londinensis* had brought him only praise.

The three principal artists responsible for making the drawings for the *Flora Londinensis* and the earlier numbers of the *Botanical Magazine* were William Kilburn, James Sowerby and Sydenham T. Edwards. William Kilburn (1741–1818) was the son of an Irish architect. He was apprenticed to a Dublin calico-printer, but on his father's death settled in London where his floral designs caught the eye of Curtis, who offered him employment. Later, however, he returned to the more lucrative world of industry. His paintings are careful and accurate; but his training had left its mark on his style, and at times his work—and in particular his original drawings, many of which can be seen at Kew—has a rather mechanical appearance. A better artist, and far more distinguished botanist, was James Sowerby (1752–1822), a member of a family which has produced many important naturalists and illustrators of natural history subjects. Sowerby studied painting at the Royal Academy Schools, and, after a brief association with the marine painter Richard Wright, turned to botanical illustration. His best work is to be found in the *English Botany* (1790–1814) which he produced in conjunction with Sir James Smith. He was no less skilful as an engraver, and the execution of many of the plates for Sibthorp's *Flora Graeca* was

PLATE 16

John Edwards made large prints for A Collection of Flowers *(London 1783–95).*
Some types are quite suited to his printmaking technique, such as the broad, smooth petals of these tulips.
Soft-ground etching plates print areas of tone, not lines;
therefore the print describes the form of flowers and leaves without cross-hatching.
Inking this tonal etching in pale green, not black,
allows for the clarity and realism of the hand-coloring.

Tulipe des Jardins.
Tulipa gesneriana L.
G. Van spaendonck del.
P. F. Le Grand sculp.

entrusted to him. The third of Curtis's protégés, Sydenham Edwards (1769?–1819), was a Welsh boy whom he discovered and trained. For twenty-eight years, these three artists were responsible for almost the whole of the illustrations for the *Botanical Magazine* which, after the death of Curtis, had been managed by his friend John Sims. In 1815, as the result of a misunderstanding, Edwards broke away and started the rival *Botanical Register*. The *Botanical Magazine*, however, was sufficiently established to survive the shock of losing in a relatively short space of time its chief artists; we shall have occasion to mention it again when we come to discuss the work of its most prolific illustrator, Walter Fitch.

From books of botanical value we now turn to some of those which appeal more particularly to the lover of decorative flower-prints. Mary Lawrance's *Roses* (1799) is the work of a quite second-rate botanical artist. Yet its plates, cunningly framed, look delightful on a wall and, to judge by the price it now commands, the volume deserves a place among the Great Flower Books. Miss Lawrance describes herself as 'Teacher of Botanical drawing etc.' One may legitimately wonder how her pupils fared, and whether she encouraged them to imitate what Dunthorne terms the 'meaningless network of lines' that pass in her own drawings for leaves. A German named Roessig, who himself published a book on roses that was far from satisfactory, condemns Miss Lawrance in words that might have been attributed to jealousy were they not undeniably true: 'Her botanical draughtsmanship is mostly inaccurate, and she rarely takes notice of the characteristic features of a rose. The lighting of the leaves is extremely bad, and only here or there one comes across one or two roses which are unmistakably identifiable'. Though John Lindley produced in 1820 a respectable *Rosarum Monographia* with adequate plates, and though in the twentieth century Alfred Parsons made for Ellen Willmott a series of delicate rose-paintings (which were sadly ill-treated by the chromolithographer), the greatest of all interpreters of the rose remains the Frenchman Redouté.

A good deal has recently been written about Thornton and his *Temple of Flora*. Indeed, all that is known of his life and of the story of the publication of his great book being readily available, it seems unnecessary to do more here than reiterate the principal facts.[14]

Robert Thornton (1768?–1837) was the son of Bonnell Thornton, a successful miscellaneous writer. The boy was destined for the Church but soon abandoned theology for medicine. The death of his father, mother and elder brother left him with a comfortable fortune which, even before he had left Cambridge, he had decided to devote to the production of a splendid volume illustrating the Linnean System of classification. *The New Illustration of the Sexual System of Linnaeus*, of which *The Temple of Flora* forms the section that here concerns us, was announced in 1797 and issued in parts between 1799

PLATE 17

Gerrit van Spaendonck pioneered the technique of color-printed stipple etching made famous by his pupil, Redouté [q.v.]. From the twenty-four prints of his Fleurs dessinées d'après nature *(Paris 1801) we feature "Tulipes des Jardins," engraved by P. F. Le Grand. Some modern experts consider this book unsurpassed in beauty, and it is the rare copy that survives complete.*

and 1807. Everything possible was done to make it the most sumptuous florilegium of all times. Competent painters such as Reinagle, Pether, Edwards and Henderson were invited to make paintings for the plates, and the Queen—'Her Gracious Majesty, the Bright Example of Conjugal Fidelity and Maternal Tenderness, Patroness of Botany and of the Fine Arts'—graciously permitted them to be dedicated to her.

The book is a bibliographer's nightmare, for hardly any two copies are quite the same. Briefly, it may be said to consist of a series of calligraphic title-pages, portraits, and allegories, followed by a set of some twenty-eight or more flower plates showing plants in their (allegedly) natural surroundings. In Thornton's own words, 'Each scene is appropriate to the subject. Thus in the night-blowing CEREUS you have the moon playing on the dimpled water, and the turret-clock points XII, the hour at night when this flower is in its full expanse . . . In the DODECATHEON, or *American* COWSLIP a sea view is given, and a vessel bearing a flag of that country . . .' Were his artists aware, we may ask, that the American cowslip does not flourish on the sea shore, and that the night-blowing cereus is unlikely to succeed in an English churchyard? Though such inaccuracies as these might (and should) have troubled Thornton, they matter little to us today. These grand, romantic engravings and Thornton's flamboyant prose we now value as evocations of a vanished age rather than as documents of scientific worth.

Fourteen of the botanical plates are the work of Peter Henderson, better known in his day as a painter of portrait and genre subjects; eleven more are by Philip Reinagle, who slaved as a copyist for Allan Ramsay before turning to original work; and two more we owe to the versatile Abraham Pether, who also added the 'moon-light' to Reinagle's night-blowing cereus. The hyacinths are by Sydenham Edwards, and the roses by Thornton himself. The mezzotint-engravers Ward, Earlom and Dunkarton, and the aquatint-engravers Stadler and Sutherland, are among those who executed the plates, which were printed in basic colours and worked over with water-colour.

Thornton had put every penny of his inheritance into his venture. He had risked all—and the Fates turned against him. He attributed his failure to the continental war, and no doubt the war was in part to blame. In an *Apology* to his clients for issuing less than half the number of plates originally intended, he wrote: 'The once *moderately rich* very justly now complain that they are exhausted through *Taxes* laid on them to pay armed men to diffuse *rapine*, *fire*, and *murder*, over *civilised* EUROPE'. He might, however, had added that changing taste had also played its part in his downfall. Without Josephine, Redouté would soon have found himself in a similar plight—and Thornton had no Josephine. The Englishman looked desperately round for a way to extricate himself. He remembered that Boydell, whose *Dramatic Works of Shakespeare* had reduced him to similar straits, had, in 1805, obtained the permission of Parliament to organize a lottery. Boydell had succeeded; he would try his luck. In 1811, permission was given for the holding of a *Royal Botanical Lottery*, with twenty-thousand tickets at two guineas each and ten thousand prizes valued at a total of £77,000. Thornton's own estimate of the worth of his prizes (which consisted of the original paintings, copies of the book, sets of plates, etc. etc.) must have been generous, since even the sale of all the tickets would only have realised £42,000. And no such success attended it. We do not know the precise extent of the disaster; we do know, however, that when

Thornton died in 1837 his family was almost destitute. But his failure is more memorable than many triumphs, and *The Temple of Flora*, after some rather surprising fluctuations, has now permanently established itself as one of the greatest prizes of collectors of fine flower-books.

In 1806, Samuel Curtis—a first cousin of William Curtis, whose only daughter he married—embarked upon a publication intended to emulate, if not to surpass, the *Temple of Flora*. *The Beauties of Flora*, for its greater rarity, has never been so well known as Thornton's book, whose fate it presumably shared. Its ten plates, which are even larger than those of the *Temple of Flora*, are mostly the work of Clara Maria Pope, who also illustrated Curtis's only other published book, *A Monograph of the Genus Camellia* (1819); there is a fine series of original paintings of peonies by her hand in the Natural History Museum, London. Mrs Pope, *née* Leigh, was first married to Francis Wheatley, creator of the *Cries of London*, and subsequently to the actor Alexander Pope. She exhibited regularly at the Royal Academy—miniatures, rustic subjects, portraits and finally groups of flowers; an ambitious example of her art is the picture, in the Sir John Soane Museum, of a bust of Shakespeare surrounded by the flowers mentioned in his plays. Except for T. Baxter's 'Tulips' which is dated 1806, and possibly the 'Ranunculuses', none of the plates of *The Beauties of Flora* was published until 1820. No doubt war was largely responsible for this delay. In general, the flowers, like those of Thornton, are shown against a landscape background (and a comparison of the more similar plates in the two books—e.g. the tulips—is interesting). The two plates of dahlias, however, have white grounds and the flowers and leaves have been treated with gum arabic to increase the richness of the colours. This technique was also used in the plates of the *Monograph on the Genus Camellia*, another immense folio with five aquatint engravings, as beautiful and as rare as its companion.

The only work that invites comparison with *The Beauties of Flora* and *The Temple of Flora* is George Brookshaw's *Pomona Britannica* (1812). Its author, who was a teacher of flower-painting and his own artist, describes his book as 'a collection of the most esteemed fruits at present cultivated in this country together with the blossoms and leaves of such as are necessary to distinguish the various sorts from each other. Selected principally from the Royal Gardens at Hampton Court and the remainder from the most celebrated gardens around London'. The ninety plates of this luscious folio are engraved in aquatint and stipple, printed in colour and finished with water-colour. In most of them, the fruit is represented against an aquatint background of dark brown which, although it is plain, suggests the drama of Thornton's and Curtis's great plates. Two hundred and fifty-six varieties of fruit are shown. Some are accompanied by their flowers; this makes it legitimate to stretch a point and include among our reproductions one of the melons, which are certainly the most decorative plates in the book.

Pomona Britannica was published at £59 18s and dedicated to George, Prince Regent. Perhaps folio fruit-books could still find purchasers in this country when folio flower-books could not. At all events, there was no need for a Royal Pomological Lottery. Indeed, in 1817 there appeared a quarto edition of *Pomona Britannica*, superior to the quarto edition of the *Temple of Flora* which had been hurriedly and carelessly produced in 1812 to provide Thornton with lottery prizes. Brookshaw followed

Erica Sebana

Franc. Bauer del. *Mackenzie sculp.*

up his successes with *A new Treatise on Flower Painting: or Every Lady her own drawing master*, and two companion volumes on birds and fruit. His method, he assures us, was so foolproof that often he could not tell his pupils' copies from his own originals; where his own flower-paintings were concerned, this was not such high commendation as he imagined.

Mrs Edward Bury (*fl.* 1831–37) of Liverpool, who made the paintings for a folio entitled *A Selection of Hexandrian Plants* (1831–34), was, on her own admission, an amateur. The plates probably owe their undoubted effectiveness to the skill of the engraver Robert Havell, for some original paintings by Mrs Bury in Major Broughton's collection would seem to confirm her amateur status. She had 'no pretension whatever', she assures us, to knowledge of her subject, and was forced into print only by the importunity of admiring friends and her own enthusiasm for 'the brilliant and fugitive beauties of a particularly splendid and elegant tribe of plants'. How fortunate she was to secure the collaboration of Havell, who played his part in such fine books as Audubon's *Birds of America* and Daniell's *Views of India*! The flowers illustrated by Mrs Bury are lilies, crinums, pancratiums and hippeastrums, grouped under the Linnean System as 'hexandrian' plants (i.e. with 6 stamens). The picture of *Amaryllis crocata* (plate 43) shows Mrs Bury at her boldest and Robert Havell at his most skilful. Yet even in this fine plate we can see that aquatint is not so well suited to botanical work as it is to architectural. Stipple and mezzotint make possible the subtle graduations of tone essential to the full interpretation of leaf and flower.

With James Bateman's *Orchidaceae of Mexico and Guatemala* (1837–41) we come to the use of a new medium eminently suited to botanical illustration—that is, lithography. Since 1812, when Ackermann published his *Series of Thirty Studies from Nature*, a number of lithographically-printed flower-books had appeared, but Bateman's atlas folio surpasses, both in beauty and in size, all that went before or came after. It contains a number of vignettes by Cruikshank, one of which caricatures its unmanageable bulk.[15]

James Bateman (1811–97) was a man of means with a mild interest in theology and a passion for horticulture. At the age of twenty-two he sent, at his own expense, a collector to Demerara and Berbice. It was, however, a merchant trading with Guatemala who obtained for him the orchids that illustrate his great folio. Since each of the forty plates cost more than £200 to produce, the book—which was published in an edition limited to one hundred copies at twelve guineas each—must have involved him in considerable personal expense. But he could well afford the loss, and there was plenty left over for the construction of his 'Chinese garden', 'Egyptian court' and 'Wellingtonia avenue' at Biddulph.

PLATE 18

Daniel Mackenzie engraved "Erica Sebana" after a painting by Franz Bauer for the latter's Delineations of Exotick Plants *(London 1796–1803). The volume illustrates varieties of heaths discovered in England's Cape colonies in Africa. Heaths also grow in England, and their distribution in the southern hemisphere greatly interested English horticulturists.*

Thirty-seven of the drawings for Bateman's folio were made by Mrs Withers or Miss Drake. The former, a teacher of botanical drawing, became 'Flower Painter in Ordinary to Queen Adelaide' and contributed to many other works of the period. Of Miss Drake, beyond her illustration to this and other books and periodicals, our knowledge is limited to the bleak fact that she lived at Turnham Green. They were good, careful, conscientious artists, as their original drawings testify; and in M. Gauci they were served by an outstanding lithographer. The hand-colouring of the plates could hardly be bettered. In every respect, *The Orchidaceae of Mexico and Guatemala* is a model of book production; and John Lindley's *Sertum Orchidaceum*, also illustrated by Miss Drake, is hardly less admirable.

The success of the *Botanical Magazine*, and a widespread interest in flowers, encouraged many nineteenth century English horticulturists to try their luck in the speculative field of the horticultural periodical. Some of these publications had long and useful lives; others were deservedly ephemeral; most of them still make delightful desultory reading, and it is much to be regretted that so many of these pleasant books have been broken up to satisfy the modern craze for floral tablemats, lampshades and the like.

A word may be said here about 'nature-printing'. As early at least as the fifteenth century, experiments had been made in taking prints from pressed leaves and flowers. At first, the technique employed was of the simplest: the specimen was inked, paper laid upon it, and pressure applied by hand. But in the nineteenth century, craftsmen of the Imperial Printing Office in Vienna discovered a method of passing the plants under pressure between a lead and a steel plate and making an electrotype from the impression made upon the lead plate. In England, Thomas Moore used this procedure for the plates of his superb *The Ferns of Great Britain and Ireland* (1857).

Another artist employed by James Bateman was Walter Hood Fitch (1817–92). Fitch, a young apprentice to a Glasgow firm of calico designers, had attracted the attention of Sir William Hooker, who then occupied the Chair of Botany in that city. When Hooker was appointed Director of Kew Gardens, he carried his protégé south with him. That was in 1841; for the next fifty years Fitch remained at Kew, and his career is inseparably associated with those of Sir William and his son Joseph. Fitch was a quick worker and extremely industrious; his published drawings reach the almost incredible total of 9960, and it is only possible here to refer to one or two of the most important works for which he provided illustrations. For forty-three years he was attached to the *Botanical Magazine*, for the greater part of which time he was its sole draughtsman and lithographer. In 1851, he illustrated Sir William Hooker's slim folio monograph on the Victoria regia waterlily (*Victoria amazonica*), which two years earlier had flowered at Chatsworth for the first time in this country. This loyal gesture (for the plant had been named in honour of the Queen) was to be rewarded later when Sir Joseph Hooker, appealing to Disraeli for a Civil List pension for Fitch, 'played upon his imperialist feelings' by showing him these big lithographs, and thus carried the day. Fitch was no less successful as the interpreter of the works of others. *The Rhododendrons of Sikkim-Himalaya* (1859–61) was illustrated with hand-coloured lithographs made after the field sketches of Sir Joseph Hooker; and in Hooker's *Illustrations of Himalayan Plants*, which appeared in 1855, Fitch adapted the drawings of Indian artists who had worked under Cathcarts's direction. At the age of sixty, Fitch began work on his last great series of lithographs, which were to form the illustrations to the *Monograph of the Genus Lilium* (1877–80) of H. J. Elwes. The supplement to this famous book, produced between 1934 and 1940, was illustrated by Miss Lilian Snelling, who proved a worthy successor to the industrious Scot.

In the words of Sir Joseph, Fitch was an 'incomparable botanical artist' with an 'unrivalled skill in seizing the natural character of a plant . . . I don't think that Fitch *could* make a mistake in his perspective and outline, not even if he tried'. Certainly his execution was brilliant; what he so often lacked was sensibility. Thus, though he always satisfies the botanist, he often leaves the artist with a sense of disappointment. His earlier drawings are tighter in handling and more detailed than his later, and by the time he came to work for Elwes he had developed a style that was almost too loose. W. B. Hemsley saw Fitch making the lily drawings direct upon the stone 'with a rapidity and dexterity that was simply marvellous', though he had, of course, made some slight preliminary sketches in pencil. His greatest gift was 'a marvellous power of visualising plants as they lived and of retaining their image in his memory'. He was the last of the giants.

We have now reached the end of our appointed journey.

But this must not lead us to suppose that the present century can show nothing comparable to the works of the past. Two important modern flower-books have already been mentioned: Ellen Willmott's *The Genus Rosa*, and Grove and Cotton's *Supplement to Elwes' A Monograph of the Genus Lilium*. The great interest in flower-books today, and the immense technical advances that have recently been made in colour reproduction, are already beginning to bear fruit in the publication both of facsimiles of the masterpieces of the past and of books illustrated by living artists. There is no lack of talent or of patronage today. Is it too much to hope that a new Golden Age of the florilegium is dawning?

THE NOTES

1. We may instance the *florilegium* or book of flower drawings by Pieter van Kouwenhoorn, now in the Lindley Library in Vincent Square. The artist was, it seems, a glass painter and the master of Gerard Dou. This little book is unsurpassed in beauty, and it is no less than astonishing that owing to the terms of its imprisonment it should be so little known. Another of these captives is the *florilegium* painted for Count Johann of Nassau by Johann Walther of Strasbourg, in two volumes, now incarcerated in the Victoria and Albert Museum. My notes on this mention the frontispiece of crown imperials and tulips growing in formal beds in front of the castle of Idstein in the Taunus mountains, near Frankfurt; the Count and his family are there, and Flora who presents them with a basket of red roses; this is followed by plates of green hellebores, crocuses, daffodils, and exquisite fritillaries. I noted a glorious drawing of crown imperials, of a double crown imperial, and one of '*Narcissus Pisanus*'; three particularly beautiful anemone drawings, and eleven tulip drawings. The third of his iris paintings I considered to be the model for all iris drawing; and later noted a splendid *Iris susiana*, a lovely martagon lily, and a beautiful foxglove; also, a madonna lily and love-in-the-mist, aquilegias, and a very fine drawing at the end, on a dark background, of candytufts and lilies. In his second volume there is a ravishing flower bouquet; a flowering aloe in a tub, more martagon lilies, a painting of strawberries, four carnation drawings; some red poppies on a dark ground, red and yellow variegated jasmine, (or is it Marvel of Peru?), ending with a section of fruits, including a lively and appetising branch or twig of currants and a thistle. These notes are given just to show the beauty of this book of drawings. Another inmate of a prison cell is the *florilegium* of Johann, Count of Dermatt, painted by Johann Simula in 1720, now languishing in the Natural History Museum in Cromwell Road. Mr Wilfrid Blunt in his *The Art of Botanical Illustration*, p. 126, says that more than a thousand flowers are represented in it, including sixty-two carnations, fifty-four tulips, forty-seven anemones, forty polyanthuses, and thirty fritillaries. He reproduces one drawing of carnations (and one from van Kouwenhoorn). The tulips and the auriculas are disappointing. On the other hand the carnations are unsurpassed in beauty, and there can be no other such collections of these Jacobean flowers. Lovely, too, are the garden poppies, the ranunculuses, and the anemones. It ends with many drawings of cactuses in china pots. On the whole, though, it is inferior to the MS. by Johann Walther in the Victoria and Albert Museum. A fourth *florilegium* by Alexander Marshal, in two volumes, wilts in more luxurious incarceration in the Royal Library at Windsor. This I have not seen; but Mr Wilfrid Blunt writes that 'the most beautiful of his paintings are those of fringed and laced pinks and carnations.' An authority on auriculas, the Rev. C. Oscar Moreton, has told me that this book of paintings is invaluable for the study and history of auriculas. Marshal is said to have perfected a secret method of painting with colours extracted from the flowers. The four *florilegia* briefly indicated here must be typical of how many others of which no more than a handful of persons is aware. But the history of human taste and skill is imperfect from that absence.

2. Madam Merian, under the influence of her friend Anna-Maria van Schuurman, joined the sect of Labadists and stayed for a long period in their headquarters in a castle in Friesland offered them by a former Governor of Surinam. He had brought back cabinets of butterflies and other insects which awoke Madam Merian's interest and determined her to go to Surinam. For a fuller account of Madam Merian herself, her friend Anna-Maria van Schuurman, and their joint hero Jean de Labadie, founder of the Labadists, see the present writer's *The Netherlands*, London, B. T. Batsford & Co., 1948, pp. 97–103.

3. Mr Wilfrid Blunt, the indispensable authority, in his *The Art of Botanical Illustration*, p. 122, and in his *Tulipomania*, King Penguin, 1950, draws attention to the Dutch tulip books, 'sale catalogues commissioned by bulb dealers to enable them to display their wares to clients when the tulips were not in season.' Eighteen such volumes have been listed, mostly, it is sad to say, by hack artists, but there is a tulip book by Judith Lyster, pupil of Frans Hals, now in a private collection, most appropriately, in Haarlem, and a *Theatrum Tuliparum* of tulips flowering in the Elector's garden in Berlin in 1647–48, now, or until lately, in the Staatsbibliothek in Berlin. An inaccessible and romantic companion to these must be the book of Turkish tulip paintings in a library at Istanbul, loc. cit. p. 164, footnote.

4. Artist of the *Fürstlische Baumeister*, Augsburg, 1711–16, an extraordinary work of the German Baroque age wherein are palaces and colonnades reaching out of sight, interiors with grand beds of state, and a rare part on gardens where Paulus Decker reveals his hand as master *in excelsis* of the clipped hornbeam hedge. A work almost on the scale of Piranesi, but with a Teutonic eye for detail which enlivens his scenes of architecture with coaches and horses and with crowds of persons.

   As for the *Nürnbergische Hesperides* (1708–14), this had an Italian predecessor, the *Hesperides* of Ferrarius (1646). Fruits and flowers tied up with ribbons as though for country dances through the streets, and surprisingly, in an age of Italian printing only otherwise enlivened by the works of Jacques Callot and of Stefano della Bella, frontispieces and a few plates by lesser masters of the Italian *seicento*.

5. A note on these volumes of Chinese Flower Paintings appears in *Journal of the Royal Horticultural Society* for June, 1953, pp. 209–13. The writer is Patrick M. Synge.

6. I have had in my possession one of the preternaturally rare striped auriculas of the seventeenth century which came to me through the plant collector Mr Kenneth W. Sanderson from, I think, an old garden in Cumberland. It was a purple striped green, and would be called, in present circumstances, an Alpine and not a stage auricula. Historical interest apart, this flower of the Jacobean age was a little dull.

7. Robert Herrick, whom it is appropriate to mention in any book concerned with flowers, published his first book of poems in 1648, when he was fifty-seven years old. The whiskered bust of this Devon clergyman, toga'd like a Roman Emperor, and with his head surrounded by a swarm of bees, appears as frontispiece and is to be counted among the delights, not only of all poetry, but of gardening, too.

8. *Chinese Painting*, by William Cohn, London, The Phaidon Press, 1948.

9. The garden hybrids of the hibiscus have been produced in such widely separated sub-tropical localities that it is doubtful if there is anywhere a complete collection of them. I have, myself, seen thirty or thirty-five varieties growing in gardens in Florida. Work has been in progress in California, the West Indies, South and East Africa, New Guinea and adjacent islands, and particularly in Hawaii, where two Scots brothers have produced many hundreds of varieties.

10. I have been sent *Empereur du Maroc*, reputedly one of the only two survivors from the golden age of the ranunculus. It was a poor thing and did not long survive.

11. S. Curtis, author and instigator of *The Beauties of Flora*, produced as well, the equally rare *A Monograph on the Genus Camellia* (1819), also with plates after Clara Maria Pope, but there are only five of them (though there are a few unpublished drawings), and it was presumably planned as a bigger work, but failed.

12. Picture-books of *ornamental* plants, as opposed to the 'herbals' which deal principally with useful plants.

13. An exclusive religious sect, founded by the French Pietist Jean de Labadie (1610–74).

14. See G. Dunthorne, *Flower and Fruit Prints of the Eighteenth and Early Nineteenth Centuries* (1938); F.M.G. Cardew, in *J. Roy. Hort. Soc.* 72: 281–85, 450–53 (1947); Wilfrid Blunt, *The Art of Botanical Illustration*, pp. 203–08 (Collins, 1950); and Geoffrey Grigson and Handasyde Buchanan, *Thornton's Temple of Flora* (Collins, 1951).

15. Mr. William Stearn tells me that this librarian's nightmare weighs 38½ lbs. μέγα βιβλίον μέγα κακόν (the greater the book, the greater the evil).

Reinagle pinxt

AN INTRODUCTION TO THE BIBLIOGRAPHY

Patrick M. Synge

MANY COLLECTORS OF OLD BOOKS with fine coloured plates value them chiefly for the decorativeness of their plates or title-pages or for their rarity and tend to disregard their intrinsic importance as books, that is, as a means of conveying information and ideas through the printed word. Nevertheless many books in this bibliography, though listed primarily for their illustrations, are still well worth reading and even re-reading and contain much that is still as valuable to botanists and gardeners as when first written. They record the introduction of new plants and reflect changing fashions in horticulture. Thus they form an endless and fascinating source of information for the student of the history of gardening, travel and even adventure. Plant-hunting has never been without its risks: many collectors died in the field and others endured great hardship in furnishing gardens and herbaria with the material so finely illustrated within these books.

The history of these great flowerbooks is indeed closely linked with the introduction of new plants in our period. The eighteenth and nineteenth centuries, when these books were produced, were times of greater personal leisure, for the wealthy at least, far more than now belongs to the majority of us, rich or poor. They were times when the flowering of a new plant was a triumph to be shown off, an exciting and sufficiently unusual event for it to be recorded in the best and most lavish manner possible. They were times also of low taxes, but plentiful and cheap labour and even cheaper coal and coke, so that stoves and warm greenhouses could be well maintained.

The Elizabethan period of travel and adventure did not bring in many new plants from overseas, although the potato is perhaps the most notable exception. The Stuart periods brought us more plants from Europe and Asia Minor. The beginning of our period in the eighteenth century and indeed for the succeeding hundred years was notable for the great number of plants which came from the Cape of Good Hope and to a lesser extent from the eastern part of North America. We had to wait nearly another hundred years for plants from the western part. Later in the century with the return of Captain Cook and Sir Joseph Banks from the Antipodes came plants from Australia and Tasmania and later New Zealand. From the middle of the nineteenth century the vogue for Cape Heaths waned, but that for warm greenhouse plants reached a greater pitch than ever before, and many treasures were brought from South America as well as from the East Indies, fascinating Orchids and pitcher plants and weird Aroids. Today the flowerbooks with fine coloured plates tend to deal with the more hardy plants such as Lilies, Irises, Roses, Paeonies and Magnolias.

PLATE 19

R. Earlom engraved this mezzotint of tulips after a painting by Philip Reinagle. It appears in The Temple of Flora *(London 1799–1807), an opulent folio produced by Robert Thornton to illustrate Linnaeus's system of plant classification. The varieties of tulips include 'Louis XVI'* (top) *and 'General Washington'* (center).

The cult of selection, often by rather rigid and artificial standards, and plant breeding in the florists' flowers has been consistent throughout our period, although the desiderata may have changed with fashion. Nor is the keenness diminished today although we cannot portray them with the magnificence of Ehret and Trew. Their great volumes, produced early in our period, show broken tulips and double hyacinths which in their day created an unprecedented enthusiasm for these plants. Alas, practically none of these old varieties can be found now, but they were very beautiful and are a distinct loss to our gardens. We know today that the breaking of tulips results from a virus infection, but still the seventeenth and eighteenth century tulips seemed to be able to live with virus and the plates are evidence of their vigour.

Andrews' books on *Heaths* were the products of a great vogue for the culture of these Cape plants at the end of the eighteenth and beginning of the nineteenth century, but now not one tenth of the species and varieties he portrays could be found in the British Isles and some may even have become extinct in South Africa. Their successful cultivation required great skill and continuous care as well as rather dry heat. Still, it is of great value and interest to have here their pictures and descriptions. The same applies to Robert Sweet's *Geraniaceae* and to a lesser extent to Redouté's *Les Roses*. Even in spite of the great enthusiasm which there is today for growing these old shrub roses, comparatively few of Redouté's roses could be readily obtained now. Of the plants figured in *Les Liliacées*, which deals mainly with wild species from the families Liliaceae, Iridaceae and Amaryllidaceae, a much larger proportion could be found again. His accurate representations remain of great value as well as beauty, and are frequently quoted in monographs dealing with the genera of these families. A few plates from these old works with their accompanying descriptions actually replace the type specimens on which certain species are based. The middle of our period from roughly 1760 to 1860 has been justly called "the golden century of botanical illustration and flowerbook production"; many people of wealth and culture were enthusiastic to sponsor and purchase finely illustrated books.

The improvements in glasshouse construction and heating in the earlier part and middle of the nineteenth century made possible the successful cultivation of plants from the warm moist tropics and naturally led to the production of fine books on orchids and other glasshouse plants, among which Camellias were then included. A few wealthy patrons of horticulture, the Horticultural Society of London, and a few enterprising nurserymen sent collectors to search the tropics for plants of beauty, as in this century they have sent them to the Himalayas and the mountains of China and the Andes in search of hardy plants. Hence such a fine work as James Bateman's *The Orchidaceae of Mexico and Guatemala*. Bateman employed the best flower painters he could find and it is doubtful whether their work could be done better today or indeed so well. He had sense of humour enough to be able to include

PLATE 20

"A Group of Auriculas" is a color-printed aquatint modeled on a painting by Peter Henderson. It illustrates Robert Thornton's The Temple of Flora *(London 1799–1807), the most romantic of English flower books. These auriculas are shown flourishing in their native Alps, but they are actually carefully bred varieties and not wildflowers.*

Henderson pinx.t
Burke & Lewis sculp.t
The Sacred Egyptian Bean
London Published Dec.r 1.1804, by D.r Thornton

even a small joke about the size of his unwieldy tome in the form of a tail-piece drawn by George Cruikshank, which shows a group of men staggering under its weight, while exultant little demons watch their plight. Some of these orchids have now been lost to cultivation in this country or are very rare indeed, but it is probable that anyone following his text could go to Guatemala and bring them back again, though it is hardly likely that he would take the book with him as part of his air baggage.

Probably our British wild flowers have never been so well portrayed and described as in James Sowerby's *English Botany* with its text by Sir James Smith, and in William Curtis's *Flora Londinensis.* They tell us much about the distribution of the plants at that time, a distribution which has sadly changed for the worse, especially with regard to some of our rarities, such as the Lady's Slipper Orchid, and also in the case of the floras, our city and urban areas. The war, however, brought into the bombed sites of London, for a short time at any rate, new and unexpected species. These books serve as a reminder to us of the continual change that is taking place both in our native and our garden floras.

Certainly the flowers of Greece and the Eastern Mediterranean region have never been better nor more extensively portrayed than in the wonderful volumes of Sibthorp and Smith's *Flora Graeca*, of which the plates by Ferdinand Bauer are masterpieces of accurate botanical drawing. After a recent visit to Greece I appreciate this work even more than before.

The middle of the nineteenth century brought the opening of Himalayan travel by botanists and Sir Joseph Hooker's name will always remain prominent among those of plant collectors and botanical travellers. His *Rhododendrons of the Sikkim-Himalaya* still remains a standard work on the subject and it is surprising to relate that no new species of Rhododendron has been discovered in Sikkim beyond those which he found and described, and what is perhaps still more astonishing, the names which he gave them have stuck, valid and undisputed by botanists, to his original plants. There are still in Scottish gardens some magnificent specimens of trees such as *Rhododendron Falconeri* grown from the seeds collected and sent home by Hooker; specimens which undoubtedly rival and probably often excel those to be found wild among the mountains. Before the publication of Hooker's book these superb plants were quite unknown in England, as was also the lovely *Magnolia Campbellii*, a flower for which it is difficult to find any rival.

At the same time, about the middle of the nineteenth century, the publication of Fitch's great plates of *Victoria regia* (now unfortunately called *V. amazonica*) introduced gardeners to one of the great marvels of the botanic world. Later in the century the publication of George Maw's *Monograph of the Genus Crocus* and Elwes's *Monograph of the Genus Lilium* gave us two standard works which have by no means been superseded even today, although the discovery of many new lilies in China and Tibet led to the magnificent supplement by Grove and Cotton, published in this century.

PLATE 21

"Sacred Egyptian Bean" was etched in aquatint and stipple by Burke and F. C. Lewis after a painting by Peter Henderson. It illustrates Robert Thornton's The Temple of Flora *(London 1799–1807), a folio which "exploits all that is most sensational in the vegetable world" (King). Printing* à la poupée *in four colors creates the translucence of the flowers.*

Iris Germanica *Iris Germanique*

P. J. Redouté pinx. de Gouy sculp.

While we must confess that some of the most important works from the botanical point of view, such as Linnaeus's *Species Plantarum*, being unillustrated and published in a small octavo size, do not come into our bibliography, there are plenty of books both finely conceived and illustrated which have played a very considerable part in the system of plant nomenclature as we know it today. In fact, owing to the strict application of the articles relative to priority of publication and rejection of homonyms in the International Code of Botanical Nomenclature, and the, to many gardeners, seemingly over-persistent activities of the strict followers of the code as they delve ever deeper and deeper into every irregularity of publication or description, the study of these beautiful books has led to the upsetting of many widely-used names in favour of older names previously published or long overlooked, perhaps because the botanists of those days were not always sufficiently familiar with the works of their predecessors.

While the binomial system of Linnaeus has remained as an established part of our present naming of plants, his sexual system of classification has long been superseded. Nevertheless from it came the inspiration for many of these fine flower books of the hundred years from 1750–1850 and frequent references to it will be found in their titles and prefaces.

The production of these books was invariably a very lengthy business and usually a very costly one. At no period has it been possible to publish cheaply books with so much hand-work and with such fine materials but yet with a comparatively limited demand. Curtis nearly beggared himself by his *Flora Londinensis*; the *Temple of Flora* ruined Thornton; the printing and colouring of the *Flora Graeca* swallowed the fortune bequeathed by Sibthorp for its publication. Consider the labour of engraving by hand all the plates and then hand colouring them, of applying mezzotint and aquatint to the same plate, as may be seen in some of the plates of *The Temple of Flora*. It was all this personal hand-work, work of a kind which no money can command today and which no machine can really equal or surpass, that gives the best of these books their extraordinary richness of texture and fineness of line. It was an age in which time did not seem to have the same importance as today, consequently so many used it so much more effectively. Twenty years was not considered too long for a man to work at a book nor did subscribers appear to become over-impatient at waiting that time for the last part. A large number of these fine flower books were issued originally to subscribers in part; even *The Temple of Flora* first appeared in this way.

Consequently when they were put together as bound volumes the dates on the title pages were often misleading and did not give accurate information about the dates of publication of the parts. In many cases, owing to one vagary or another, they do not even represent the actual date of publication of the completed volume. Since the actual date of publication, even of individual pages and plates, may

PLATE 22

Les liliacées *(Paris 1802–16) by Pierre Joseph Redouté set a new standard of excellence for botanical books. "Iris Germanica" exemplifies the realism achieved through the use of color-printed stipple etching. Redouté groomed an atelier of thirty artisans in this technique, of which De Gouy was one of the eighteen who made the 486 prints of this work.*

be important for establishing priority of nomenclature, many botanists have taken considerable trouble and displayed unusual powers of detection in unearthing and recording these dates, often in scientific journals which are not always very familiar to the collector. Therefore it has been thought worthwhile to give many references to such papers and dates of issue in this bibliography.

Thus many of these fine flower books were originally issued in a paper bound form nearer to a magnificent periodical than to the form of a book. The differentiation between the two was far less than today; the present bibliography contains a separate section for periodicals, but in it are some borderline cases which may be equally well thought of as books.

Enormous care was also given to the fine periodicals such as *The Botanical Magazine* and they have played a part both in horticulture and botany that few other works have done. They were finely conceived and often finely executed. Therefore the best of them are included in this bibliography in their own section.

The Botanical Magazine merits a special eulogy. The first part appeared in 1787. It is unique in that it is still being published, and still going strong, although alas in numbers sold not nearly so strong as when it was first started by William Curtis to provide some of the 'pudding' which his *Flora Londinensis* had so sadly failed to ensure for him. It is the oldest scientific magazine in the world with coloured plates which is still being published. In these volumes, now 170 in number, there is abundant material for a history of plant collecting over the last two centuries. Each plate with its accompanying text takes us on a miniature journey of exploration and it is a continuous story as much alive now as when the magazine was founded. The first plate was *Iris persica*, now alas practically unknown in English gardens, but it is still, to my mind, one of the most delightful flower pictures ever issued. Its editors have been among the greatest botanists of the day and its association with Kew is a very old one. *The Botanical Magazine* is indeed a project which well merits support and a complete set is a most valued and prized possession. Of all the works surveyed in this bibliography I think that it is the one which I prefer to possess, since not only are many of the plates beautiful in themselves, but the volumes are an endless treasure house of wonderful plants. Of very special distinction and a joy to handle also is the fine set on large paper, printed so that the double plates are not folded, which belonged first to Sir William Hooker, then to his son Sir Joseph Hooker and to the latter's son-in-law, Sir William Thiselton-Dyer, then to H. J. Elwes, and is now in the Lindley Library, London.

A word should also be said about the *Transactions* of the Horticultural Society of London published between 1807 and 1848 in two series. As examples of fine periodical production or even as book production, they reached a very high, indeed almost a luxurious, standard, especially in their plates, while on turning the pages one realizes what a large number of interesting and extremely practical papers are contained therein.

It is one of the aims of this bibliography of great flower books to present them as living works, useful as well as beautiful, and to persuade our readers to regard them as complete entities, worthy to be preserved in their full glory as fine books and not to be treated merely as a source of curious pictures for lampshades or table mats.

The interest of the fine flower book is well maintained throughout our period, but it is sad to see how much of its distinction and beauty has in so many cases been lost before the end of our two centuries. The advent of colour-printing by chromolithography, while cheapening the production considerably, did not by any means produce such a fine book. The finest papers are also to be found in the earlier periods.

To previous bibliographers, Georg August Pritzel, Gordon Dunthorne and Claus Nissen, whose bibliography is probably the most complete yet published of all flower books, the editor and the compilers of this bibliography owe much. Our thanks are also due to Major the Hon. Henry Broughton, whose library includes some flower books not represented in any of our public collections.

Lilium Superbum *Lis Superbe*

P. J. Redouté pinx. de Gouy sculp.

EXPLANATORY NOTES

Difficulties of definition immediately confront us when deciding which books to list in a bibliography primarily concerned with great flower books. Between the really fine flower book of folio size, handsomely printed on firm rag-made paper, and profusely illustrated with coloured plates by a master of botanical illustration—such a book, for example, as Sibthorp and Smith's *Flora Graeca* (1806–40)—and the humble octavo textbook with a few simple diagrams, there is every intermediate. Neither size nor colouring by itself furnishes adequate criteria for decision. A badly drawn or crudely coloured illustration is none the better for occupying a folio plate, while exquisitely detailed and accurate plates may adorn a work no more than 5 inches by 3½ inches in size, as Sturm's *Deutschlands Flora in Abbildungen* (1798–1862) proves. Moreover to list only works with coloured plates would exclude many important books with monochrome engravings and lithographs by A. Riocreux, d'Apreval, A. Faguet and others of the nineteenth century French school of botanical artists, and these for delicacy and accuracy can hardly be excelled.

To provide a standard a great flower book may be defined as one in which text and plates alike are of high quality, a book which gives pleasure in all its parts to a connoisseur. Many such books are among those listed below, but the bibliography is not confined to them: too high a standard, too rigid and narrow a definition, would limit its usefulness by excluding many books of lasting value and interest from both aesthetic and botanical standpoints. In scope the following bibliography comes between Dunthorne's *Flower and Fruit Prints of the 18th and early 19th Centuries* (1938), with its detailed collations and strong emphasis on artistic appeal, and Nissen's more comprehensive and scientific treatise *Die botanische Buchillustration* (1951–52). Books devoted solely to ferns, mosses, fungi and algae have been excluded, not from lack of appreciation of their beauty and importance, but because they merit separate treatment. In this connexion it is worthy of note that 'the finest plates ever published in a botanical work', to quote T. G. Hill's commendation of Riocreux's illustrations in Thuret's *Études Phycologiques*, depict algae. Probably only with ferns and other plants notable for the intricacy of their foliage does this technique give pleasing results; an impression made direct from a specimen of the plant itself records beautifully and precisely the tracery of veins in leaves, but the general effect, as in Kniphof's *Botanica in Originali* (1747) tends to be dismal and is necessarily flat. Only a few outstanding books on fruit have been included; fruit books as a whole also merit separate treatment. An excellent survey of these by that scholarly pomologist and epicure E. A. Bunyard will be found in the *Journal of the Royal Horticultural Society* 40:414–49 (1915). Books to teach young ladies the art of flower painting have for the most part been excluded along with sentimental or light-hearted little books on the language of flowers. Encyclopaedias, travel books, reports of surveys and voyages and similar works in which illustrations of plants, sometimes indeed of excellent quality, mingle with others of animals,

---

PLATE 23

"Lilium Superbum" is Plate 103 of Les liliacées *(Paris 1802–16) by artist Pierre Joseph Redouté. This costly book illustrated the flowers of Empress Josephine's garden in an edition of 218 eight-volume sets. To insure the financial success of her favorite artist, Josephine prompted the French government to buy a large part of the edition for use as gifts of state.*

Agapanthus Umbellatus *Agapanthe en Ombelle.*

P. J. Redouté pinx. Dequoy sculp. Tassaert direx.

scenery, minerals, etc., have mostly received the same harsh treatment, although these have been excluded with greater reluctance on account of their value to science. Certain entomological works, in which appear some charming and accurate portraits of plants, are also omitted, since in them the flowers and foliage are represented, admittedly sometimes gnawed and frayed, as the host plants of insects; examples are Moses Harris, *The Aurelian or Natural History of English Insects, namely Moths and Butterflies, together with Plants on which they feed* (London, 1766) and its French edition, *L'Aurelien* (London, 1791), and John Curtis, *British Entomology* (8 vols., London, 1823–40). Exception has been made for Maria Sibylla Merian's *Metamorphosis Insectorum Surinamensium* which brought the wonders of tropical American vegetation, no less than of its insect fauna, to the eyes of eighteenth century Europe; and also for Smith Abbot's *Lepidopterous Insects of Georgia.*

Omitted too are the great ornithological works of Gould and Audubon, in which splendid illustrations of plants serve as backgrounds to illustrations of birds; remove the birds from 180 plates in Audubon's *Birds of America* (1827–38) and his great bird book would become a great flower book! The choice of entries for a list of flower books is necessarily a subjective matter. Some books have been included because, although their plates are not artistically outstanding, they give the first representations of species. The reader's pardon is asked for such errors, if errors they be, of inclusion as of omission. The compilers' aim being to base entries on direct examination, the bibliography has been compiled chiefly from copies available in the major botanical and horticultural libraries of the London area, namely the Lindley Library of the Royal Horticultural Society, the herbarium library of the Royal Botanic Gardens, Kew, and the library of the Department of Botany, British Museum (Natural History). In addition, permission was granted to one of the authors of the bibliography to examine the Library of Major the Hon. Henry Broughton and collate a number of books nowhere else available.

SIZE

The size of a page is determined by the size of the original sheet of paper as it comes from the printing press, and the manner in which it is later folded and trimmed. Thus a sheet 50 cm. high, 70 cm. broad makes four *folio* pages 50 cm. high, 35 cm. broad when folded once but makes eight *quarto* (4to) pages 35 cm. high, 25 cm. broad when folded twice; according to the manner of folding it will make sixteen *octavo* (8vo) pages 25 cm. high, 17½ cm. broad or twenty-four duodecimo (12mo) pages 16⅔ cm. high, 17½ cm. broad or 24 cm. high, 12 cm. broad when folded four times. Sheets vary, however, considerably in size and thus the terms folio, quarto, octavo and duodecimo provide only rough and general indications of the size of pages and plates. Hence they are supplemented below by measurements in centimetres, the height first, then the width, taken mostly from copies in the libraries mentioned. Some

PLATE 24
Les liliacées *(Paris 1802–16) by Pierre Joseph Redouté depicts the magnificent variety of flowers from the lily and other families in Empress Josephine's vast gardens. "Agapanthus umbellatus" is a stipple etching inked in four colors and hand-colored. It demonstrates the artist's strong and graceful draftsmanship, and the success of this printmaking technique for showing the luster and modeling of leaves.*

PLATE 67.

Silver Rock Mellon.

Painted & Pub. by the Author as the Act directs. 1812.

of these have undoubtedly been cut down by their binders and their measurements may not apply exactly to all copies.

PLACE OF PUBLICATION

The place of publication of the earlier works with a Latin text is often stated in Latin form on the title-page. Thus books printed at Paris may bear the imprint *Lutetiae, Lutetiae Parisiorum* or *Parisiis*, because modern Paris covers the area where in Julius Ceasar's time the Gallic tribe Parisii had their town Lutetia. Similarly *Lugduni Batavorum* may not be easily recognized as referring to Leiden, *Olyssipone* to Lisbon, *Sebastianopolis* to Rio de Janeiro, *Tiguri* to Zürich, *Oeniponte* to Innsbruck. The locative case is employed rather than the nominative. Such Latinized place-names are followed by their modern equivalent in brackets.

DATE OF PUBLICATION

The date of publication is accepted as that on the title-page unless there exists evidence to the contrary. Unfortunately the title-page date is often inaccurate and misleading. Since the scientific nomenclature of plants, as of animals, rests upon priority of publication, it is important to ascertain the correct date, for when there exist two or more names for the same plant, the correct name is usually the one published first. Many fine flowerbooks originally appeared in parts. Sometimes the title-page came out in the first part, in which event much of the work will be of later date. Thus the title-page of Mikan's *Delectus Florae et Faunae Brasiliensis* bears the date '1820' but the text contains references to the years 1821 and 1822. Investigation reveals that this book appeared in four parts (part 1 in 1820, part 2 in 1822, part 3 in 1823, part 4 in 1825). Sometimes none of the work appeared in the year stated; for example, unfortunate circumstances delayed until 1834 or 1835 the publication of Velloso's *Florae Fluminensis Icones* which is dated '1827' on its title-page. An exactly opposite situation arises when the title-page bears the date of the last part. Elwes's *Monograph of the Genus Lilium* has the date '1880' on the title-page but only the sixth and final part appeared then; publication of the first part was in 1877. Sometimes the title-page gives inclusive dates, e.g. Webb and Berthelot's *Phytographia Canariensis* is dated '1836–1850', but these lack precision. To establish the contents and the dates of publication of these works is often a difficult business, involving examination of watermarks and long and tedious search through contemporary periodicals, booksellers' catalogues, library accession registers and the correspondence and diaries of botanists, followed by a critical evaluation of the evidence, which is often scanty and sometimes conflicting and hence demands considerable knowledge of book production and publishing.

Original parts in their wrappers as published are invaluable for such research but are now hard

PLATE 25

The complement to Curtis's and Thornton's memorable flower books [q.v.] is George Brookshaw's Pomona britannica *(London 1804–12). "Silver-Rock Mellon" is one of the delectable English fruits among the book's ninety color-printed aquatint etchings. The seeds and skin are beautifully expressed by aquatint alone without the use of lines.*

to find. An incomplete copy of a fine flower book having the wrappers preserved may thus be scientifically more important than a sumptuously bound perfect copy from which the wrappers have been discarded. From William Roscoe's correspondence with Sir James Smith between 1823 and 1829 the course of publication of Roscoe's *Monandrian Plants* becomes clear; unfortunately we lack information about the contents of the parts. The old archives of the Hon. East India Company reveal when the four last parts of Roxburgh's *Plants of the Coast of Coromandel* were issued but are silent about their contents. Anyone possessing parts in wrappers of these and similar chronologically enigmatic works would do well to inform the Librarian of the British Museum (Natural History) London, so that they can be recorded. Nevertheless the manner of publication of many important books has now been satisfactorily elucidated. The bibliography accordingly gives references to papers where information of this kind may be found and also makes available much information not hitherto published. Dates given in square brackets, e.g. 1820 [–25] come from sources other than the work itself. Dates in round brackets, e.g. 1820 (–25) have been deduced from statements in the work.

Many of the prints in English flower books published between 1734 and 1845 will be found to have a dated statement at the foot such as "Publ. by S. Curtis, Walworth May 1 1811" on plate 1331 of Curtis's *Botanical Magazine* or "Pubd. as the Act Directs June 1st 1803 by H. Andrews, Knightsbridge", on plate 305 of Andrews' *Botanist's Repository*. This statement is sometimes very faint, needing a lens for examination, and has sometimes been cut away by the bookbinder. When published it had the strictly utilitarian purpose of securing copyright under the British print copyright Acts of 1734, 1766 and 1777, but it now yields valuable information on the date of publication of the plate. The 1734 Act resulted from an appeal to Parliament by William Hogarth and "*Hogarth's Act*" is indeed a simpler title for it than its official designation, *An Act for the Encouragement of the Arts of Designing, Engraving and Etching Historical and other Prints*. This Act gave to inventors, designers and engravers of prints the sole right of printing the same for the term of fourteen years "to commence from the Day of first publishing thereof which shall be truly engraved with the Name of the Proprietor on each Plate and printed on every such Print or Prints". The 1766 Act extended protection for 28 years. The 1777 Act maintained this protection and prohibited the copying of a print in whole or part without the written and witnessed consent of the owner. The date had, of course, to be engraved on the copper plate ahead of the day of publication of the print but it was in the publisher's interest that the two should agree and such dates on prints are usually reliable.

One of the aims of this bibliography is to enable collectors and librarians to ascertain whether their copies are complete or not. For this reason the *number of plates* has been counted in copies available and checked against statements in other bibliographies, notably these of Pritzel and Dunthorne. It is not feasible, however, in a general work to deal with variations of such notoriously complex works as Lambert's *Description of the Genus Pinus* and Thornton's *Temple of Flora*, of which it appears that no two copies are exactly the same. For these reference must be made to detailed accounts elsewhere, e.g. to Renkema and Ardagh's paper on Lambert's *Pinus* in the *Journal of the Linnean Society* 48: 439–66 (1930) and to Handasyde Buchanan's bibliographical notes in Grigson and Buchanan's *Thornton's Temple of Flora*, 13–19 (1951).

The earlier works having plates with an engraved or lithographed outline frequently exist in both hand-coloured and uncoloured states. Sometimes, as with Curtis's *Flora Londinensis*, uncoloured states are extremely rare; sometimes, as with Allioni's *Flora Pedemontana*, the coloured state is extremely rare. Hence some books described here as having coloured plates will also be found in an uncoloured state, while it is possible that a few works not mentioned as having coloured plates may exist in a coloured state unknown to the compilers.

Customary abbreviations have been employed in citing periodicals, e.g. *J.* for *Journal*, *Bull.* for *Bulletin*. Thus:

J. Arnold Arb.	= Journal of the Arnold Arboretum of Harvard University, Cambridge, Mass.
J. Bot. (London)	= Journal of Botany British and Foreign, London.
J. Soc. Bibl. Nat. Hist.	= Journal of the Society for the Bibliography of Natural History, London.
Bull. Torrey Bot. Club	= Bulletin of the Torrey Botanical Club, New York.
Contrib. Gray Herb.	= Contributions from the Gray Herbarium of Harvard University, Cambridge, Mass.

Throughout the Bibliography, asterisks indicate that a plate number is repeated once (*), or twice (**). For example, the entry for Reichenbach & Others (1834–1914) includes a collation of the plates. This collation also features the numeral suffix "bis" to indicate repeated plate numbers. Authors resorted to these bibliographic conventions when supplementing their publication with additional plates. Such plates often feature species variations made known to the author during the course of publication.

Entries ending with the initials H.B. are those compiled by Handasyde Buchanan.

PLATE XXVI.
Painted & Published as the Act directs by the Author G. Brookshaw. April. 1806.

THE BIBLIOGRAPHY

W. T. Stearn, Sabine Wilson and Handasyde Buchanan

AITON, WILLIAM

HORTUS KEWENSIS; OR, A CATALOGUE OF THE PLANTS CULTIVATED IN THE ROYAL BOTANIC GARDEN AT KEW.
3 vols 8vo (20½ cm × 13 cm). London 1789.
13 engraved plates (of which Nos. 2, 4, 6, 10–12, are much larger and folded in) by Fr. Bauer, Ehret, J. F. Miller, Nodder, J. Sowerby, engraved by McKenzie.

ALLEN, JOHN FISK

VICTORIA REGIA; OR THE GREAT WATER LILY OF AMERICA . . .
Folio (69 cm × 54½ cm). Boston, Mass. 1854.
6 chromolithographed plates by William Sharp, lithographed by Sharp and Son.

ALLIONI, CARLO

FLORA PEDEMONTANA SIVE ENUMERATIO METHODICA STIRPIUM INDIGENARUM PEDEMONTII.
3 vols of which Vol. 3 is Atlas of plates (FLORAE PEDEMONTANAE ICONES).
Folio (39½ cm × 26 cm). Augustae Taurinorum (Turin) 1785.
92 uncoloured plates by F. Peiroleri, engraved by P. Peiroleri. Coloured copies are extremely rare.

ALYON, PIERRE PHILIPPE

COURS DE BOTANIQUE POUR SERVIR À L'ÉDUCATION DES ENFANS DE S.A. SÉRÉNISSIME MONSEIGNEUR LE DUC D'ORLÉANS, OÙ L'ON A RASSEMBLÉ LES PLANTES INDIGÈNES ET EXOTIQUES EMPLOYÉES DANS LES ARTS ET DANS LA MÉDECINE.
8 parts Folio (46 cm × 30 cm). Paris [1787–88].
103 hand-coloured plates drawn and engraved by Jean Aubry.

ANDREWS, HENRY C.

COLOURED ENGRAVINGS OF HEATHS. THE DRAWINGS TAKEN FROM LIVING PLANTS ONLY . . .
4 vols Folio (42½ cm × 28 cm). London (1794–) 1802–09 (–1830).
288 hand-coloured engraved plates all by H. C. Andrews.
The plates in this and other works by Andrews are rather crude and stylised. W.B.

ANDREWS, HENRY C. See Periodicals.

ANDREWS, HENRY C.

GERANIUMS, OR A MONOGRAPH OF THE GENUS GERANIUM; CONTAINING COLOURED FIGURES OF ALL THE KNOWN SPECIES AND NUMEROUS BEAUTIFUL VARIETIES . . .
2 vols 4to (30 cm × 24 cm). London 1805.
124 unnumbered hand-coloured engraved plates by H. C. Andrews.

ANDREWS, HENRY C.

ROSES, OR A MONOGRAPH OF THE GENUS ROSA; CONTAINING COLOURED FIGURES OF ALL THE KNOWN SPECIES AND BEAUTIFUL VARIETIES.
2 vols 4to (29 cm × 23½ cm). London 1805–1828.
Frontispiece and 129 unnumbered hand-coloured engraved plates by H. C. Andrews. For dates of publication, see E. M. Tucker in *J. Arnold Arb.* 18: 258–260 (1937).

ANDREWS, HENRY C.

THE HEATHERY; OR A MONOGRAPH OF THE GENUS ERICA, CONTAINING COLOURED ENGRAVINGS.
6 vols 8vo (approx. 24 cm × 15 cm). London 1804–12.
300 unnumbered hand-coloured engraved plates by H. C. Andrews.
Ed. 2, 'corrected and enlarged', London 1845.
6 vols 8vo (approx. 24 cm × 15 cm).

ANDREWS, JAMES

FLORA'S GEMS, OR THE TREASURES OF THE PARTERRE. I WREATH AND II BOUQUETS DRAWN AND COLOURED FROM NATURE. WITH POETICAL ILLUSTRATIONS BY L.A. TWAMLEY.
Folio (37 cm × 27½ cm). London [c. 1830–37].
12 coloured engraved plates by J. Andrews.
A skilful artist, too much of whose time was wasted upon sentimental volumes of this kind. W.B.

PLATE 26

Dark, squared vignettes distinguish the plates of George Brookshaw's Pomona britannica *(London 1804–12) from all other fruit prints. The velvety texture of these peaches is color-printed, and the fruits are finished with delicate hand coloring.*

ANDREWS, JAMES

LESSONS IN FLOWER PAINTING. A SERIES OF EASY AND PROGRESSIVE STUDIES DRAWN AND COLOURED AFTER NATURE.
Folio (27 cm × 18½ cm; landscape shape). London 1835.
24 engraved plates by J. Andrews, of which 12 are plain and 12 coloured.

ANDREWS, JAMES

THE PARTERRE OR BEAUTIES OF FLORA.
Folio (37 cm × 27 cm). London 1842.
12 hand-coloured lithographs.
A companion volume to FLORA'S GEMS. H.B.

ANONYMOUS (variously attributed to Duke and Carwitham)

THE COMPLEAT FLORIST.
8vo (22 cm × 13 cm). London 1747.
Coloured frontispiece, coloured title-page and 100 hand-coloured engravings.
2nd edition 1795 with second title-page after plate 50.
Printed by J. Duke. Sold by J. Robinson. Frontispiece signed 'J. Carwitham sculp'. H.B.

ANONYMOUS

NEDERLANDSCH BLOEMWERK.
4to (28½ cm × 20½ cm). Amsteldam (Amsterdam) 1794.
Title-page and 53 hand-coloured engraved plates by P. T. van Brussel, engraved by H. L. Myling.

ANONYMOUS

DARSTELLUNG VORZÜGLICHEN AUSLÄNDISCHEN BÄUME UND GESTRÄUCHE WELCHE IN DEUTSCHLAND . . .
Vol. I (all published) 4to (29 cm × 22 cm). Tubingen (J. G. Cottaischen Buchhandlung) 1769.
60 hand-coloured engravings. H.B.

ANONYMOUS

LA CORBEILLE DE FLEURS, OUVRAGE DE BOTANIQUE ET DE LITTÉRATURE.
8vo (21 cm × 13 cm). Paris 1807.
24 plates printed in colour and finished by hand.
A companion volume LE PANIER DE FRUITS was published at the same time also with 24 plates. H.B.

ANONYMOUS [E. and S. M. FITTON]

CONVERSATIONS ON BOTANY.
Small 8vo. London 1817.
20 hand-coloured engravings. H.B.

ANONYMOUS

ICONES PLANTARUM SPONTE CHINA NASCENTIUM, E BIBLIOTHECA BRAAMIANA EXCERPTAE.
See KER, C.H.B.

ANONYMOUS

(TEN) LITHOGRAPHIC COLOURED FLOWERS BY A LADY.
2 vols Folio (35 cm × 16 cm). Edinburgh n.d. (1826).
40 hand-coloured lithographs.
No title page. Title 'Ten Lithographic coloured Flowers' is taken from the first part. H.B.

ANONYMOUS

ROSES ET ROSIERS PAR DES HORTICULTEURS.
4to (27 cm × 18 cm). Paris n.d. (1840).
48 hand-coloured lithographs. H.B.

ANONYMOUS

THE ORNAMENTAL FLOWER GARDEN AND SHRUBBERY, CONTAINING COLOURED FIGURES AND DESCRIPTIONS OF . . . FLOWERING PLANTS AND SHRUBS CULTIVATED IN GREAT BRITAIN, SELECTED FROM THE WORKS OF JOHN LINDLEY, R. SWEET, D. DON, ETC.
4 vols 8vo (24½ cm × 6 cm). London 1852–54.
Plates and text taken from the *Botanical Register* and Sweet's *British Flower Garden* with Latin text translated into English.

ANONYMOUS

CHOIX DES PLUS BELLES ROSES.
Folio (27 cm × 17 cm). Paris (Dusacq) n.d. [1845–1854].
60 hand-coloured lithographs from paintings by Annica Bricogne.
Text by Victor Paquet. H.B.

ANONYMOUS

DE NEDERLANDSCHE BOOMGAARD BESCHREVEN EN UITGEGEVEN . . . MET AFBEELDINGEN NAAR DE NATUUR.
2 parts 4to (30 cm × 24 cm). Groningen [1864–]1868.
124 chromolithographed plates. Part I. Nos. 1–60; Part II. Nos. 1–36, 1–8, 1–12, 1–4, 1–4, by S. Berghuis, lithographed by G. Severeyns, 2 title-pages by O. Erelman.

ANONYMOUS (H.C.W.)

WILD FLOWERS OF SWITZERLAND OR A YEAR AMONGST THE FLOWERS OF THE ALPS.
Large 4to. London 1883.
16 chromolithographs by H. C. Ward, lithographed by F. Frick. H.B.

ANTOINE, FRANZ, *the younger*

PHYTO-ICONOGRAPHIE DER BROMELIACEEN DES K.K. HOFBURG-GARTENS IN WIEN.
4to text (33½ cm × 26 cm) with Folio Atlas (65 cm × 48 cm). Wien [Vienna] 1884.
Title-page and 35 plates by F. Antoine, lithographed by C. Höller, 30 plates partly hand-coloured.

ARCHER, THOMAS CROXEN

PROFITABLE PLANTS; A DESCRIPTION OF THE PRINCIPAL ARTICLES OF VEGETABLE ORIGIN USED FOR FOOD . . . ETC.
8vo (16 cm × 12½ cm). London [1865].
20 hand-coloured lithographed plates by W. H. Fitch.
A re-issue of *Popular Economic Botany*, London 1853.

BAKER, J. G. See BURBIDGE, F. W., 1875.

BANKS, SIR JOSEPH, & SOLANDER, DANIEL CARL

ILLUSTRATIONS OF THE BOTANY OF CAPTAIN COOK'S VOYAGE ROUND THE WORLD IN H.M.S. ENDEAVOUR IN 1768–71. WITH DETERMINATIONS BY JAMES BRITTEN.
3 vols Folio (48½ cm × 32 cm). London 1900–05.
320 lithographed plates (1–318, 41A, 45A) by T. Burgis, J. Cleveley jr, J. F. Miller and J. Miller, F. P. Nodder (after the sketches by S. Parkinson), originally engraved by D. Mackenzie, G. Sibelius, R. Blyth, Goldar, G. Smith, White.
3 maps
Although published just outside our period (part 1, pp. 1–31, pls. 1–100 in 1900; part 2, pp. 35–75, pls. 101–243 in 1901; part 3, pp. 77–102, pls. 244–318 in 1905) this is included here because it should have been published some 125 years earlier! The original drawings and sketches were made on Captain Cooks' first voyage by Sydney Parkinson, who died at sea in 1771, then completed in England by J. Cleveley, J. F. Miller and others. Engraving of the plates began in 1772; unfortunately their publication was delayed and, after the death of Banks's companion Solander in 1782, was postponed indefinitely and did not take place until the above issue in 1900–1905 of lithographed copies. They are of special interest as being at the time of their preparation the first representations of Australasian plants, but they do not do justice to the beauty of the original drawings now in the Department of Botany, British Museum (Natural History).

BARKER-WEBB, PHILIP & BERTHELOT, SABIN

HISTOIRE NATURELLES DES ISLES CANARIES.
3 vols. Large 4to (35 cm × 25 cm). Atlas Folio (53 cm × 34 cm). Paris 1835–50.
Originally issued in 106 parts.
Tome 3, Part 2: PHYTOGEOGRAPHIA CANARIENSIS. About 287 lithographed black and white plates by various artists with 20 plates and maps in atlas, of which 9 illustrate scenery. The plates are numbered up to 252 but some numbers have separate additional plates numbered A or B. Numerous text figures by various artists. For a detailed account of this work and dates of publications of individual plates see W. T. Stearn in *Journ. Soc. for the Bibliography of Nat. Hist.* 1:49–64 (1937).

BARKER-WEBB, PHILIP

OTIA HISPANICA, SEU DELECTUS PLANTARUM RARIORUM AUT NONDUM RITE NOTARUM PER HISPANIAS SPONTE NASCENTIUM.
Folio (44½ cm × 30 cm). Parisiis & Londini (Paris & London) 1839.
6 engraved plates (tabs 1–10, with tabs 6–10 on one plate) by A. Chazal, A. Riocreux, and Mme. Spach, engraved by Mougeot and E. Taillant. Pentas 1, pp. 1–18, tabs. 1–10 (uncoloured); Pentas 2, pp. 9–15, tabs 6–10 (all on one plate, coloured).

BARKER-WEBB, PHILIP

OTIA HISPANICA SEU DELECTUS PLANTARUM AUT NONDUM RITE NOTARUM, RARIORUM PER HISPANIAS SPONTE NASCENTIUM.
Folio (36 cm × 26 cm). Parisiis (Paris) 1853.
42 plates (tabs. 1–45, 36a, with 6–10 on one plate) by A. Chazal, J. Desne, Félix, Fitch, Mme. Spach, A. Riocreux and Vaillant, engraved by E. Taillant, Picart and Mougeot. This includes all of the 1839 edition on pp. 1–17 (text by Montague is altered) and its pls. 1–10.

BARLA, JEAN BAPTISTE

FLORE ILLUSTRÉE DE NICE ET DES ALPES MARITIMES. ICONOGRAPHIE DES ORCHIDÉES.
4to (36 cm × 27 cm). Nice 1868.
63 coloured plates by J. B. Barla and V. Fossat. lithographed by E. Carlin, V. Fossat, A. Giletta, the Lea Brothers, Lubain.

BARRATTE, JEAN FRANÇOIS GUSTAVE. See BONNET, E. 1895 and COSSON, E. S., 1882.

BARTHOLOMEW, VALENTINE

A SELECTION OF FLOWERS, ADAPTED PRINCIPALLY FOR STUDENTS.
Small Folio (30 cm × 25 cm). London 1822.
24 hand-coloured lithographs. *Very fine lithography by C. Hullmandel.* H.B.

BARTHOLOMEW, VALENTINE

GROUPS OF FLOWERS.
Small Folio (30 cm × 25 cm). London n.d. [1823].
6 hand-coloured lithographs.
A supplement to his Selection of Flowers (above). Have been found bound together as one book. H.B.

BARTON, BENJAMIN SMITH

ELEMENTS OF BOTANY, OR OUTLINES OF NATURAL HISTORY OF VEGETABLES.
8vo (approx 23 cm × 14 cm). Philadelphia 1803.
30 hand-coloured plates. Revised and corrected London 1804.
Ed. 2. Corrected and enlarged, Philadelphia, 1812–14.
Ed. 3. 1827. New ed. 1836.

P. Bessa pinx.t
Clement sculp.t

BARTON, WILLIAM PAUL CRILLON
VEGETABLE MATERIA MEDICA OF THE UNITED STATES, OR MEDICAL BOTANY . . .
2 vols 4to (27 cm × 21 cm). Philadelphia 1817–18.
50 hand-coloured plates by W.P.C. Barton, engraved by Tanner, Vallance, Kearney and Co., J. Boyd, and J. G. Warnicke.

BARTON, WILLIAM PAUL CRILLON
A FLORA OF NORTH AMERICA, ILLUSTRATED BY COLOURED FIGURES DRAWN FROM NATURE.
3 vols 4to (27½ cm × 22 cm). Philadelphia 1821–23.
106 hand-coloured plates, drawn by W.P.C. Barton, engraved by G. B. Ellis, F. Kearny, C. Thiebout, J. Boyd, J. Drayton. J. L. Frederick, C. Goodman, Jacob J. Plocher, coloured by W.P.C. Barton and family. In some copies there is no plate numbered 88 but two plates are then numbered 92 (*Cephalanthus occidentalis* and *Sarothra hypericoides* with *Malaxis ophioglossoides*).

BATEMAN, JAMES
THE ORCHIDACEAE OF MEXICO AND GUATEMALA.
Folio (73 cm × 53 cm). London [1837–] 1843.
40 hand-coloured plates by Miss Drake, Miss Jane Edwards, Holden, Mrs Withers, lithographed by M. Gauci. Vignettes by Lady Jane Walsh, Lady Grey of Groby, Mrs R. Wilbaham, George Cruikshank (p. 8 and under pl. 9), J. Brandard, R. Branston, T. P. Woox, E. Landells, C. U. Skinnor, G. Ackermann, engraved by R. Branston and E. Landells, are an entertaining feature of this massive book; it is indeed the only botanical work with illustrations by George Cruikshank.
The largest, heaviest, but also probably the finest orchid book ever issued. The text is sound also for growers. W.B.

BATEMAN, JAMES
A MONOGRAPH OF ODONTOGLOSSUM.
Folio (55 cm × 38 cm). London. [1864–74].
30 hand-coloured plates, drawn and lithographed by W. H. Fitch.
For dates of publication, see W. T. Stearn in *Flora Malesiana* I.4: clxviii (1954).
Fitch here shows incredible ability in dealing with complicated botanical specimens. W.B.

BATEMAN, JAMES
A SECOND CENTURY OF ORCHIDACEOUS PLANTS, SELECTED FROM . . . CURTIS' BOTANICAL MAGAZINE SINCE THE ISSUE OF THE 'FIRST CENTURY'.
4to (31 cm × 24½ cm). London 1867.
100 hand-coloured plates drawn and lithographed by W. H. Fitch.
A continuation of Hooker, W. J., *A Century of Orchidaceous Plants*.

BATSCH, AUGUST JOHANN GEORG KARL
DER GEÖFFNETE BLUMENGARTEN.
8vo Weimar 1798.
100 hand-coloured engravings taken from early numbers of Curtis' Botanical Magazine, H.B.

BAUER, FERDINAND LUKAS
ILLUSTRATIONES FLORAE NOVAE HOLLANDIAE, SIVE ICONES GENERUM, QUAE IN PRODROMO FLORAE NOVAE HOLLANDIAE ET INSULAE VAN DIEMEN DESCRIPSIT ROBERTUS BROWN.
3 parts Folio (51 cm × 34 cm). London [1806–] 1813.
15 hand-coloured plates, drawn, engraved and coloured by Ferdinand Bauer. A 16th plate (*Lambertia formosa*) in Dept. Botany, British Museum (Natural History) probably unpublished.
A rare and very beautiful work of which probably less than 50 copies (coloured and uncoloured) were issued. *Abandoned because of the exhausting work of acting as his own engraver, Bauer being unwilling to entrust it to others.* W.B.

BAUER, FRANZ ANDREAS
DELINEATIONS OF EXOTICK PLANTS CULTIVATED IN THE ROYAL GARDENS AT KEW DRAWN AND COLOURED AND THE BOTANICAL CHARACTERS DISPLAYED ACCORDING TO THE LINNEAN SYSTEM.
Folio (about 60 cm × 48 cm). London [1796–] 1803.
30 unnumbered hand-coloured plates of Ericaceae drawn and coloured by Franz Bauer, engraved by Mackenzie, Ferdinand Bauer and J. Basire.
The finest engravings after the drawings of the incomparable Franz Bauer. W.B.

BAUER, FRANZ ANDREAS
STRELITZIA DEPICTA, OR COLOURED FIGURES OF THE KNOWN SPECIES OF THE GENUS STRELITZIA FROM THE DRAWINGS IN THE BANKSIAN LIBRARY.
Folio (55½ cm × 44½ cm). London 1818.
11 hand-coloured plates by Franz Bauer, lithographed and coloured by Harris and Moser.

PLATE 27
"Alcea rosea" embellishes Pancrace Bessa's Fleurs et fruits *(Paris 1808), a portfolio of twenty-four prints. Both Bessa and Redouté studied under van Spaendonck and used his medium of color-printed stipple etching to reproduce watercolors. In fact, they used the same etchers, the artisans ultimately responsible for the success of a print.*

BAUER, FRANZ ANDREAS

ILLUSTRATIONS OF ORCHIDACEOUS PLANTS.
2 Parts Folio (27 cm × 36½ cm). London 1830–38.
35 hand-coloured engraved plates drawn by Franz Bauer, engraved by M. Gauci.

BAUMANN, CARL & BAUMANN, NAPOLÉON.

BOLLWEILERER CAMELLIEN-SAMMLUNG.
4 Parts Folio (33½ cm × 26 cm). Bollweiler 1828–35.
49 hand-coloured lithographed plates (part 1, pls. 1–12; part 2, pls. 13–24; part 3, pls. 25–37; part 4, pls. 38–49) by C. A. and Napoléon Baumann, lithographed by Engelmann et Cie. and G. Bruckert et Cie.
French Edition: Collection des Camellias élevés à Bollwiller. 1835.

BAXTER, WILLIAM

BRITISH PHAENOGAMOUS BOTANY, OR FIGURES AND DESCRIPTIONS OF THE GENERA OF BRITISH FLOWERING PLANTS.
6 vols 8vo (21 cm × 13½ cm). Oxford [1832–] 1834–43.
509 hand-coloured engravings drawn by W. A. Delamotte, C. Mathews, Isaac Russell, H. Bidwell, Miss Saunders, W. Willis, and engraved by C. Mathews, J. Whessell and W. Willis.
For a history of this work, see A. H. Church, 'Baxter's British Phaenogamous Botany', *J. Bot. (London)* 57:58–63 (1919).
Original issue in parts was regarded as the *first edition*; completed sets of volumes as a *second edition*, and a reprint in 1856 as a *third edition*; but text and plates are the same in all.

BEDDOME, RICHARD HENRY

ICONES PLANTARUM INDIAE ORIENTALIS, OR PLATES AND DESCRIPTIONS OF NEW AND RARE PLANTS FROM SOUTHERN INDIA AND CEYLON.
1 vol. 4to (30 cm × 23 cm). Madras & London, 1868–74.
300 plates by Alwis and Govindo, lithographed by Barren and Co., and Dumphy.

BEDDOME, RICHARD HENRY

THE FLORA SYLVATICA FOR SOUTHERN INDIA: CONTAINING QUARTO PLATES OF ALL THE PRINCIPAL TIMBER TREES IN SOUTHERN INDIA AND CEYLON.
2 vols 4to (29½ cm × 23½ cm). Madras [1869–74].
359 lithographed plates (1–230, 231–330, 1–29) by Govindo, Alwis, W. Fitch, lithographed by Dumphy.
For dates of publication see W. T. Stearn in *Flora Malesiana* I. 4: clxix (1954).

BÉLANGER, CHARLES

VOYAGE AUX INDES ORIENTALES, PAR LE NORD DE L'EUROPE, LES PROVINCES DU CAUCASE, LA GEORGIE, L'ARMENIE ET LA PERSE . . . PENDANT 1825–29. BOTANIQUE. ATLAS.
2 parts Text 8vo (27 cm × 21 cm). Atlas 4to (27 cm × 21 cm). Paris [1846].
31 engraved plates by Bory de St. Vincent and E. Delile, 3 plates partly coloured.
The text deals only with Cryptogams; that for Phanerogams was never published.

BELLERMANN, JOHANN BARTHOLOMAUS

ABBILDUNGEN ZUM KABINET DER VORZÜGLICHSTEN IN- UND AUSLANDISCHEN HOLZARTEN NEBST DEREN BESCHREIBUNG.
Folio (35½ cm × 23 cm). Erfurt 1788.
73 hand-coloured engraved plates (Nos. 1–72 and 1 double plate of analyses) by J. B. Bellermann.

BENNETT, ARTHUR. See FRYER, ALBERT [1898–] 1915.

BENNETT, JOHN JOSEPH, & BROWN, ROBERT

PLANTAE JAVANICAE RARIORES DESCRIPTAE ICONIBUSQUE ILLUSTRATAE, QUAS IN INSULA JAVA, ANNIS 1802–18, LEGIT ET INVESTIGAVIT THOMAS HORSFIELD . . .
4to (36 cm × 26½ cm). Londini (London). 1838–52.
50 hand-coloured engraved plates by C. and J. Curtis engraved by J. Curtis and E. Weddell.
According to W. T. Stearn in *Flora Malesiana* I. 4: clxx (1954), pp. 1–104, pls. 1–24 were published in July 1838; pp. 105–196, pls. 25–40 in May 1840; pp. 197–238, pls. 41–45 in Nov. 1845; pp. 239–258, pls. 46–50, map in May 1852.

BENTHAM, GEORGE

THE BOTANY OF THE VOYAGE OF H.M.S. SULPHUR UNDER THE COMMAND OF CAPT. SIR E. BELCHER . . . DURING THE YEARS 1836–42, EDITED AND SUPERINTENDED BY RICHARD BRINSLEY HINDS . . . THE BOTANICAL DESCRIPTIONS BY G. BENTHAM.
4to (30 cm × 23½ cm.). London. 1844 [–1846].
60 uncoloured plates drawn and lithographed by Miss Drake.
For dates of publication, see W. T. Stearn in *Flora Malesiana* I. 4: clxx (1954).

BENTHAM, GEORGE

HANDBOOK OF THE BRITISH FLORA. Ed. 2
2 vols 8vo (20 cm × 13½ cm). London. 1865.
1295 wood-engravings by W. H. Fitch, engraved by O. Jewitt: issued uncoloured but sometimes carefully hand-coloured by amateur botanists. Ed. 3. 1873.
Actually a reissue of the 2nd edition. The first edition

of 1858 had no illustrations. In 1880 the wood-engravings were separated from the text, 11 new engravings being then added, and were issued thereafter as a separate volume under the title *Illustrations of the British Flora* by Fitch, W. H., with additions by Smith, W. C.

BENTLEY, ROBERT, & TRIMEN, HENRY

MEDICINAL PLANTS, BEING DESCRIPTIONS WITH ORIGINAL FIGURES OF THE PRINCIPAL PLANTS EMPLOYED IN MEDICINE . . .

4 vols 8vo. (23 cm × 16½ cm). London [1875–] 1880.

306 hand-coloured plates drawn and lithographed by D. Blair.

For dates of publication, see B. L. Burtt in *Notes R. Bot. Gard. Edinburgh* 21: 157–162 (1953).

BERGERET, JEAN PIERRE

PHYTONOMATOTECHNIE UNIVERSELLE, C'EST À DIRE L'ART DE DONNER AUX PLANTES DES NOMS TIRÉS DE LEUR CHARACTERES . . .

3 vols Folio (40½ cm × 26 cm). Paris 1783–84.

328 unnumbered unsigned hand-coloured plates drawn and engraved by Poisson.

BERGHUIS, S.

DE NEDERLANDSCHE BOOMGAARD . . . MET AFBEELDINGEN NAAR DE NATUUR

2 vols 4to (30 cm × 24 cm). Groningen 1864–1868.

2 chromolithographed title-pages by Erelman and 124 chromolithographed plates, nos. 1–60, 1–36, 1–8, 1–12, 1–4, 1–4, by S. Berghuis, chromolithographed by Severeyns.

BERLÈSE, LORENZO

ICONOGRAPHIE DU GENRE CAMELLIA OU DESCRIPTION ET FIGURES DES CAMELLIA LES PLUS BEAUX ET LES PLUS RARES PEINTS D'APRÈS NATURE . . .

3 vols Folio (37 cm × 28½ cm). Paris [1839–] 1841–43.

300 partly hand-coloured plates by J. J. Jung, engraved by A. Duménil, Gabriel, Oudet.

Plate 227 is called in error *Camellia heteropetala rubra*; it should be *C. heteropetala alba*.

BERTHELOT, SABIN. See under BARKER-WEBB, P. 1835–50.

BESSA, PANCRACE

FLEURS ET FRUITS GRAVÉS ET COLORIÉS SUR LES PEINTURES AQUARELLES FAITES D'APRÈS NATURE

Folio Paris 1808.

24 hand-coloured engravings.

Very well done. H.B.

BESSA, PANCRACE

FLORE DES JARDINIERS, AMATEURS ET MANUFACTURIERS D'APRÈS LES DESSINS DE BESSA. EXTRAITS DE L'HERBIER DE L'AMATEUR DE A. DRAPIEZ.

4 vols 4to. Paris 1836.

389 coloured plates by Bessa.

Bessa was a good artist who wasted too much of his time on petty works. W.B.

BICKNELL, CLARENCE

FLOWERING PLANTS AND FERNS OF THE RIVIERA AND NEIGHBOURING MOUNTAINS.

4to (27½ cm × 17½ cm). London 1885.

82 chromolithographed plates by C. Bicknell, lithographed by West, Newman & Co.

BLACKWELL, ELIZABETH

A CURIOUS HERBAL, CONTAINING FIVE HUNDRED CUTS OF THE MOST USEFUL PLANTS WHICH ARE NOW USED IN THE PRACTICE OF PHYSICK. ENGRAVED ON FOLIO COPPER PLATES AFTER DRAWINGS TAKEN FROM THE LIFE . . .

2 vols. Folio (36 cm × 23 cm). London 1737 [–39].

500 hand-coloured plates drawn and engraved by Elizabeth Blackwell.

New eds. 1739 and 1751.

Compiled by Elizabeth Blackwell to make money enough to get her husband out of a debtor's prison. Not very good engravings. W.B.

BLACKWELL, ELIZABETH

HERBARIUM BLACKWELLIANUM EMENDATUM ET AUCTUM, ID EST ELISABETHAE BLACKWELL COLLECTIO STIRPIUM QUAE IN PHARMACOPOLIIS AD MEDICUM USUM ASSERVANTUR . . . CUM PRAEFATIONE C.J. TREW.

6 vols. Folio (36 cm × 23 cm). Norimbergae (Nuremberg) 1757–73.

615 hand-coloured plates engraved by N. F. Eisenberger.

Vol. 1 (Cent. I) 100 plates. Vol. 2 (Cent. II) 100 plates. Vol. 3 (Cent. III) 101 plates. There are two plates numbered 269 of *Dracontium*. Vol. 4 (Cent IV) 102 plates: pl. 322a, *Cannabis foemina*, pl. 322b, *Cannabis mas*, pls. 341a and 341b, *Corallium rubrum*. Vol. 5 (Cent. V) 102 plates: pls. 497a and 497b, *Nymphaea lutea*; pls. 498a and 498b, *Nymphaea alba*. Vol. 6 (Cent VI) A hand-coloured title-page showing *Viscum Mancanillae*, and 110 plates: pls. 522a and 522b, *Cucurbita lagenaria*; pls. 523a and 523b, *Solanum tuberosum*; pls. 533a and 533b, *Bryonia alba*; pls. 536a and 536b, *Lupulus humulus*; pls. 539a and 539b, *Momordica balsamina*; pls. 547a and 547b, *Mays frumentum turcicum*; pl. 573a, *Cicuta viennensis* and *C. terrestris major*, pl. 573b, *Cicuta viennensis*, pls. 574a, 574b, and 574c, *Cicuta aquatica*; pls. 600a and 600b, *Rhabarbarum sinense*.

11e Cahier
Pl. 42.
J.L. Prevost pinx.
A Paris chez Vilquin, Md d'Estampes, grande cour du Palais du Tribunat No 20
Ruotte sculp.

BLANCO, MANUEL

FLORA DE FILIPINAS . . . ADICIONADA CON EL MANUSCRITO INÉDITO DEL FR. IGNACIO MERCADO, LAS OBRAS DEL FR. ANTONIO LLANOS Y DE UN APÉNDICE . . . GRAN EDICION, HECHA BAJO LA DIRECCION CIENTIFICA DEL FR. ANDRES NAVES (Y CELESTINO FERNANDEZ-VILLAR).

4 vols, Appendix and Atlas. Folio (43 cm × 30 cm). Manila 1877–80.

477 (or more) coloured lithographed plates (of which 8 are unnumbered) by C. Arguelles, F. Domingo, Juan, Reg. and Rosendo Garcia, L. Guerrero, F. Martinez and M. Zaragoza, lithographed by Oppel, M. Perez and C. Verdaguer.

The Kew copy has the following note by R. A. Rolfe on the number of plates: 'Taking the numerical sequence . . . the following numbers are blank: 2, 16, 61, 65, 67, 77, 92, 101, 103, 107, 123, 169, 186 and 325. And the following numbers are in duplicate: 43, 73, 86, 94, 100, 124, 131, 138, 167, 175, 210, 226, 368, 402, 404, 405, 414, 415, 425, 426, 427, 428 and 429. Actual number of plates published 477.468 (no. of last plate) + 23 (in duplicate) — 14 (numbers blank) = 477. (R. A. Rolfe).'

According to Nissen the plates 2–325 listed above which are lacking in most copies are present in the Berlin Staatsbibliothek copy.

BLUME, CARL LUDWIG

FLORA JAVAE, NEC NON INSULARUM ADJACENTIUM . . . ADJUTORE J. B. FISCHER.

3 vols Folio (43½ cm × 27 cm). Bruxellis (Brussels) 1828 [–51].

238 engraved and lithographed plates (of which 225 are coloured) by Arckenhausen, Bick, Latour, Sixtus, J. Vivien and Wild, engraved by W. Engels, lithographed by C. Hohe, Hütz, G. Severeyns, G. Simonau, Sixtus and Vivien.

For dates of publication, see B. H. Danser in *Blumea* 3: 203–211 (1939), M. J. van Steenis-Kruseman and W. T. Stearn in *Flora Malesiana* I. 4: clxx (1954).

It consists of the following, each botanical family having its own pagination and plate-numbering.

Rhizantheae: 6 (double) lithographed plates, of which 3 coloured, 1828.

Chlorantheae: 2 engraved and lithographed coloured plates 1829.

Dipterocarpeae: 6 lithographed, coloured plates of which 4 double 1829.

Filices: 1–94 engraved and lithographed, coloured plates (Nos. 18 and 81 double) 1829–47.

Cupuliferae: 24 engraved and lithographed coloured plates 1829.

Juglandeae: 4 hand-coloured plates 1 uncoloured plate 1829.

Balsamifluae: 1 hand coloured-plate, 1 uncoloured plate 1829.

Myricae: 1 coloured plate 1829.

Magnoliae: 12 plates of which 8 and 12 uncoloured and 9 and 10 double 1829.

Schizandreae: 5 coloured plates 1830.

Anonaceae: 53 plates (6 uncoloured, i.e. nos. 13, 14, 25, 31, 36, 52) 1830.

Loranthaceae: 28 coloured plates (except No. 23 which is uncoloured) 1830–51.

BLUME, CARL LUDWIG

FLORA JAVAE ET INSULARUM ADJACENTIUM. NOVAE SERIES . . . VOL. I: ORCHIDEAE.

Folio (45 cm × 29 cm). Amstelodami (Amsterdam) 1858 [–59].

70 lithographed plates (pls. 1–66, 9b–d, 12b) of which 56 are coloured and pl. 26 is a double plate, by Th. Bik, Blume, Gordon, Latour, van Raalten and A. J. Wendel, lithographed by P. Lauters and G. Severeyns.

BLUME, CARL LUDWIG

RUMPHIA, SIVE COMMENTATIONES BOTANICAE IMPRIMIS DE PLANTIS INDIAE ORIENTALIS, TUM PENITUS INCOGNITIS TUM QUAE IN LIBRIS RHEEDII, RUMPHII, ROXBURGHII, WALLICHII ALIORUM RECENSENTUR.

4 vols Folio (43 cm × 27 cm). Lugduni-Batavorum (Leyden), & Amstelodami (Amsterdam) 1835–48.

3 portraits and 210 partly lithographed, partly engraved plates (pls. 1–200, 157b, 163b, 167b, 172b, 172c, 176b, 176c, 178b, 178c, 200b, of which most are hand-coloured) by Arckenhausen, Berghaus, Bick, Blume, Decaisne, Gordon, Latour, Lauters and Payen, lithographed by A. Henry and Cohen, P. Lauters and G. Severeyns.

For dates of publication, see M. J. van Steenis-Kruseman and W. T. Stearn in *Flora Malesiana* I. 4: ccxviii (1954).

BOIS, DÉSIRÉ GEORGE JEAN MARIE

ATLAS DES PLANTES DE JARDINS ET D'APPARTEMENTS EXOTIQUES ET EUROPÉENNES, ACCOMPAGNÉ D'UN TEXTE EXPLICATIVE . . .

PLATE 28

Jean Louis Prévost drafted Collection des fleurs et fruits *(Paris 1805) to supply decorative artists with motifs. Printmaker Louis Ruotte transcribed the artist's beautiful renderings into stipple etching. These poppies were printed in seven intense colors of ink, and lightly finished by hand with watercolor.*

Pl. 15. | *4.e Cahier*

A Paris, chez Vilquin, M.d d'Estampes, grande cour du Palais du Tribunat, N.o 20.

J. L. Prevost inv. | *L. C. Ruotte Sc*

3 vols 8vo (23 cm × 15½ cm). Paris [1891–] 1896.
320 chromolithographed plates by B. Hering.
Some of these plates were also used in Step, E., *Favourite Flowers of Garden and Greenhouse* (1896–97).

BOISSIER, EDMOND PIERRE
VOYAGE BOTANIQUE DANS LE MIDI DE L'ESPAGNE PENDANT L'ANNÉE 1837.
2 vols and Atlas, 4to (35 cm × 27 cm). Paris 1839–45.
206 hand-coloured engraved plates (pls. 1–181, 1a, 4a, 6a, 9a, 14a, 14b, 26a, 40a, 64a, 80a, 84a, 85a, 92a, 94a, 98a, 102a, 108a, 113a, 115a, 118a, 122a, 123a, 125a, 126a, 132a) by Heyland and A. Riocreux, engraved by Borromée and Heyland. One coloured double-plate 'Tableau Synoptique'. Published in 22 parts.

BOISSIER, EDMOND PIERRE
ICONES EUPHORBIARUM, OU FIGURES DE 122 ESPÈCES DU GENRE EUPHORBIA . . .
Folio (41½ cm × 32 cm). Bâle & Paris 1866.
122 uncoloured lithographed plates (1–120, 31bis, 40bis) by Heyland, lithographed by Duc. Corraterie and Mercier et Cie.

BOLUS, HARRY
ICONES ORCHIDEARUM AUSTRO-AFRICANARUM EXTRA-TROPICARUM; OR, FIGURES WITH DESCRIPTIONS OF EXTRA-TROPICAL SOUTH AFRICAN ORCHIDS.
3 vols 8vo (25 cm × 15 cm). London 1893–1913.
300 lithographed plates, most of them partly colour-printed, by H. Bolus and F. Bolus.

BONAFOUS, MATHIEU
HISTOIRE NATURELLE, AGRICOLE ET ÉCONOMIQUE DU MAÏS.
Folio Paris, Turin 1836.
5 uncoloured and 14 coloured engravings on steel, of which one coloured plate by Redouté.

BONELLI, GIORGIO
HORTUS ROMANUS.
8 vols Folio (53 cm × 38 cm). Romae 1772–93.
800 hand-coloured unsigned engraved plates by Cesare Ubertini and L. and C. Sabatti, engraved by Magdalena Bouchard, also 5 portraits and 1 plan.

BONNET, EDMOND, & BARRATTE, JEAN FRANÇOIS GUSTAVE
ILLUSTRATIONS DES ESPÈCES NOUVELLES, RARES OU CRITIQUES DE PHANÉROGAMES DE LA TUNISIE (EXPLORATION SCIENTIFIQUE DE LA TUNISIE. BOTANIQUE).
Folio (36 cm × 27 cm). Paris 1895.
15 uncoloured plates (pls. 6–20) drawn and lithographed by C. Cuisin and Madame B. Herincq.

BONPLAND, AIMÉ JACQUES ALEXANDRE
DESCRIPTIONS DES PLANTES RARES CULTIVÉES À MALMAISON ET À NAVARRE.
Folio (52 cm × 36 cm). Paris. [1812–] 1813 [–17].
64 uncoloured plates, 54 by Redouté, 9 by Bessa and one unsigned, engraved by Bouquet, Bessin, Coutant, De Gouy, Monsaldi, Plée fils and Véran.
For dates of publication, see W. T. Stearn in *J. Arnold Arb.* 23. 110 (1942).

BONPLAND, AIMÉ JACQUES ALEXANDRE
VOYAGE DE HUMBOLDT ET BONPLAND. 6me PARTIE, BOTANIQUE, Sect. I. PLANTES ÉQUINOXIALES . . . (PLANTAE AEQUINOCTIALES . . .).
2 vols. Folio (51 cm × 34½ cm). Paris 1805–13 [–17].
143 hand-coloured engraved plates (pls. 1–140, 1b, 2b, 30b) by Turpin, engraved by Sellier.
For dates of publication, see C. D. Sherborn in *J. Bot. (London)* 39: 203 (1901). This is part of Humboldt & Bonpland's *Voyage Botanique*.

BONPLAND, AIMÉ JACQUES ALEXANDRE
VOYAGE DE HUMBOLDT ET BONPLAND. 6me PARTIE, BOTANIQUE, Sect. II. MONOGRAPHIE DES MÉLASTOMACÉES.
2 vols 1 (Mélastomes, Melastomae); 2 (Rhexies, Rhexiae).
Folio (51 cm × 34 cm). Paris (1806–) 1816–23.
120 hand-coloured engraved plates (1–60 in each volume) by Poiteau and Turpin, with the dissections of *Rhexia Hilariana* (Vol. 2, pl. 56) by Kunth, engraved by Bouquet and Ruote.
Each volume has three title-pages differently worded; hence the work may be cited as *Voy. Humboldt & Bonpl., Bot. Melast.* (see above) or as *Monographie des Mélastomacées* or *Monographia Melastomacearum*. L. C. Richard helped Bonpland in the preparation of the first volume (*Mélastomes*) and C. S. Kunth completed the second volume with the help of Auguste de Saint-Hilaire after Bonpland's departure for South America. The work was issued in 24 parts, Kunth editing the last five at Humboldt's request.

PLATE 29
In Collection des fleurs et fruits *(Paris 1805),*
Jean Louis Prévost reproduced his flowers with the technique favored by the incomparable Redouté.
Printmaker Louis Ruotte has etched this hollyhock
in the stipple manner. Inking with colored pigments creates an image suffused with
the brilliant color of actual petals and leaves.

For dates of publication see C. D. Sherborn & B. B. Woodward in *J. Bot. (London)* 39: 203 (1901), *Flora Malesiana* 1. 4: clxxxix (1954).

BOOTH, WILLIAM BEATTIE

ILLUSTRATIONS AND DESCRIPTIONS OF . . . THE NATURAL ORDER CAMELLIEAE AND OF THE VARIETIES OF CAMELLIA JAPONICA CULTIVATED IN THE GARDENS OF GREAT BRITAIN.

Vol. 1. Folio (38 cm × 27 cm). London [1830–] 1831.
40 hand-coloured plates (36 engravings and 4 lithographs) by Alfred Chandler, engraved by S. Watts and Weddell.
For a note on this work, see W. T. Stearn in R. Hort. Soc., *Camellias and Magnolias, Report of the Conference* 127 (1950). Parts 1–6, pls. 1–24 published in 1830, parts 7–10, pls. 25–40 in 1831. In 1942 six parts of Vol. 2 part 1, containing pls. 41–44, dated '1837', were found among books belonging to a descendant of the Chandler family but seem never to have been published. Vol. 1 exists in three states, i.e. with uncoloured plates, coloured plates, and more highly finished coloured plates on large paper.

BORY DE SAINT-VINCENT, JEAN BAPTISTE GEORGE MARCELLIN

NOUVELLE FLORE DU PÉLOPONNÈSE ET DES CYCLADES, ENTIÈREMENT REVUE CORRIGÉE ET AUGMENTÉE PAR CHAUBARD POUR LES PHANÉROGAMES—ET BORY DE SAINT-VINCENT POUR LES CRYPTOGAMES ETC.

Folio (51 cm × 34 cm). Paris & Strasbourg 1838.
42 plates (41 engraved and one lithographed) by Bory de St. Vincent, Borromée, A. Brocq, Chaubard and Delile, engraved by Breton, G. Coignet, Duménil, François, Georger, E. Taillant, A. Tardieu.
Plates 34 and 41 completely and plates 37, 38, and 40 partly hand-coloured.

BOWLES, CARINGTON. See SAYER, R., 1760.

BOWLES, JOHN & CARINGTON

MONTHLY FLORIST.

Folio (46 cm × 28 cm). London, Carington Bowles. n.d. [1750].
A copy of Furber's Twelve Months of Flowers (Q.V.). Coloured title-page and 12 hand-coloured engravings of groups of flowers. H.B.

BRADLEY, RICHARD

THE HISTORY OF SUCCULENT PLANTS, CONTAINING THE ALOES, FICOIDS (OR FIG-MARYGOLDS), TORCH THISTLES, MELON THISTLES . . . (HISTORIA PLANTARUM SUCCULENTARIUM . . .)

5 parts. 4to (22 cm × 17 cm). London 1716–27.
49 hand-coloured plates (pls. 1–50, with 26 and 27 on same page) by R. Bradley, engraved by John Clark, H. Hulsbergh, J. Pine and Sturt.
Ed. 2. London 1734.
For a note on this work, see G. D. Rowley, 'Richard Bradley and his History of Succulent Plants', *Cactus and Succ. J. Gt. Brit.* 16: 30–31, 54–55, 78–81 (1954).
Poor plates but important botanically and cited by later authors. W.B.

[BRADLEY, RICHARD]

THE FLOWER-GARDEN DISPLAY'D, IN ABOVE FOUR HUNDRED CURIOUS REPRESENTATIONS OF THE MOST BEAUTIFUL FLOWERS; REGULARLY DISPOS'D . . . ON COPPER-PLATES FROM THE DESIGNS OF MR. FURBER, AND OTHERS.

4to (25½ cm × 19 cm). London 1730.
Ornamental title-page and 12 hand-coloured engraved plates reduced from those of Furber's *Twelve Months of Flowers* and thus based on paintings by P. Casteels, engraved by J. Smith. These reduced copies (though of little merit) seem to have been very popular, no comparable work of higher quality being available at the time, and in two different issues sponsored by two groups of London book- and print-sellers (Robert Sayer, John King and Philip Overton; R. Montagu, J. Brindley and C. Corbett). For a note on this work, see R. P. Brotherston in *Gard. Chron. III.* 47:33 (15 Jan. 1910).
Ed. 2, TO WHICH IS ADDED, A FLOWER-GARDEN FOR GENTLEMEN AND LADIES . . . ALSO A METHOD OF RAISING SALLETINGS, CUCUMBERS, MELONS ETC. AT ANY TIME OF THE YEAR, AS IT IS NOW PRACTISED BY SIR THOMAS MORE, BART.
4to (27 cm × 19 cm). London 1734. Plates as above.

BREDA, JACOB GIJSBERT SAMUEL VAN

GENERA ET SPECIES ORCHIDEARUM ET ASCLEPIADEARUM QUAS IN ITINERE PER INSULUM JAVAM . . . COLLEGERUNT H. KUHL ET J. C. VAN HASSELT.

Vol. 1 Folio (56 cm × 40½ cm). Gandavi (Ghent) 1827 [i.e. 1828–29].
15 hand-coloured plates by P. Bik, T. Bik, F. de Keghel, G. L. Keultjens, G. van Raalten, A. Steijaert, lithographed by Kierdorff, F. de Keghel.
According to H.C.D. de Wit in *Flora Malesiana Bull.* no. 6: 165–167 (1950) part 1 was issued in 1828, part 2 & 3 in 1829.

BREYN, JAKOB

PRODROMI FASCICULI RARIORUM PLANTARUM PRIMUS ET SECONDUS . . . QUIBUS PRAEMITTUNTUR VITA ET EFFIGIES AUCTORIS, CURA ET STUDIO, JOHANNIS PHILIPPI BREYNII FILII. HUJUS AD CALCEM ANNECTITUR DISSERTATIO BOTANICO-MEDICA DE RADICE GIN-SEM SEU NISI ET HERBA ACMELLA.

4to (28 cm × 22½ cm). Gedani (Danzig) 1739.

33 uncoloured plates (3 unnumbered) by Duplon, Mylius, A. Stech, engraved by I. Saal, J. P. Kilian, Veenhuysen. Portrait engraved by G. P. Busch.

BROOKSHAW, GEORGE
POMONA BRITANNICA, OR, A COLLECTION OF THE MOST ESTEEMED FRUITS AT PRESENT CULTIVATED IN THIS COUNTRY . . . SELECTED PRINCIPALLY FROM THE ROYAL GARDENS AT HAMPTON COURT . . . ACCURATELY DRAWN AND COLOURED FROM NATURE, WITH FULL DESCRIPTIONS . . .
Folio (57 cm × 45 cm). London [1804–] 1812.
90 aquatint plates (pls. 1–93, with pls. 39, 42, 46 missing) by G. Brookshaw, engraved by R. Brookshaw and H. Merke.
Another edition 2 vols. 4to (33 cm × 26 cm). London [1816–] 1817.
60 engraved plates partly hand-coloured, partly colour-printed. In the 4to edition 'Great Britain' is substituted for 'this country' in the title.

BROOKSHAW, GEORGE. See Periodicals.

BROWN, JOHN EDNIE
THE FOREST FLORA OF SOUTH AUSTRALIA
9 parts. Folio (55 cm × 43 cm). Adelaide [1882–90].
45 unnumbered chromolithographed plates by H. B., C. H., R.C.F. (under the direction of J. E. Brown), lithographed by H. F. Leader, E. Spiller.

BROWN, ROBERT. See BENNETT, J. J. 1838.

BROWNE, PATRICK
THE CIVIL AND NATURAL HISTORY OF JAMAICA.
Folio (36½ cm × 24 cm). London 1756.
49 uncoloured plates (of which pls. 1–38 are botanical) by G. D. Ehret, engraved by R. Benning, Edwards and Darley, F. Garden, J. Noual, F. Patton, H. Roberts. 1 map.
Ed. 2. London, 1789. The text of this edition agrees with that of the first edition but has four additional indexes; presumably the original copper plates were not available for all the plates have been re-engraved, and are consequently reversed, with the names of the artist and engravers removed and with Linnaean botanical names added; they are coarser than the original plates. Although aesthetically dull these plates are valuable for the study of West Indian botany.
Ehret at his least interesting. W.B.

BRUGIÈRE, J. G., & BORY DE SAINT-VINCENT, JEAN BAPTISTE GEORGE MARCELLIN
TABLEAU ENCYCLOPÉDIQUE ET MÉTHODIQUE DES TROIS RÈGNES DE LA NATURE. VERS, COQUILLES, MOLLUSQUES ET POLYPIERS.
3 vols 4to. Paris 1791–1827.
With 488 uncoloured plates, of which 3 signed: P.J.R.

BUCHOZ, PIERRE JOSEPH
LES DONS MERVEILLEUX ET DIVERSEMENT COLORIÉS DE LA NATURE DANS LA RÈGNE VÉGÉTALE.
2 vols. Folio (44 cm × 28 cm). Paris 1779–1783.
200 hand-coloured engravings. H.B.

BUCHOZ, PIERRE JOSEPH
PLANTES, NOUVELLEMENT DÉCOUVERTES, RÉCEMMENT DÉNOMMEÉS ET CLASSÉES, REPRÉSENTÉES EN GRAVURES, AVEC LEURS DESCRIPTIONS . . .
Folio (41½ cm × 26½ cm). Paris 1779–1784.
50 unsigned plates, engraved by C. Fessard.

BUCHOZ, PIERRE JOSEPH
HERBIER OU COLLECTION DES PLANTES MÉDICINALES DE LA CHINE D'APRÈS UN MANUSCRIT PEINT ET UNIQUE QUI SE TROUVE DANS LA BIBLIOTHÈQUE DE L'EMPEREUR DE LA CHINE
Folio (45 cm × 27 cm). Paris 1781.
100 hand-coloured engravings. H.B.

BUCHOZ, PIERRE JOSEPH
LE JARDIN D'EDEN, LE PARADIS TERRESTRE RENOUVELLÉ DANS LE JARDIN DE LA REINE À TRIANON.
Folio (45 cm × 28 cm). Paris 1783.
200 hand-coloured engravings. H.B.

BUCHOZ, PIERRE JOSEPH
LE GRAND JARDIN DE L'UNIVERS, OÙ SE TROUVENT COLORIÉES LES PLANTES LES PLUS BELLES, LES PLUS CURIEUSES ET LES PLUS RARES DES QUATRE PARTIES DE LA TERRE . . .
2 vols Folio (41 cm × 26½ cm). Paris 1785–1791.
200 hand-coloured engraved plates (No. clvii 'Melle Surrugue del.', No. clviii signed 'Melle Surrugue Laine (l'aîné?) del'., and nos. lxxxix–xcv, xcvii–ci, cx–cxvi with Chinese signatures (?).

BULLIARD, PIERRE
FLORA PARISIENSIS, OU DESCRIPTIONS ET FIGURES DES PLANTES, QUI CROISSENT AUX ENVIRONS DE PARIS . . .
6 vols 8vo (20 cm × 11½ cm). Paris 1776–83.
640 unnumbered hand-coloured engraved plates by Bulliard.

BULLIARD, PIERRE
HERBIER DE LA FRANCE, OU COLLECTION COMPLETTE DES PLANTES INDIGÈNES DE CE ROYAUME.
13 vols Folio (33 cm × 24 cm). Paris 1780 [–1808].
602 coloured engraved plates, drawn and colour-printed by Bulliard.
Published in 150 parts, each with 4 plates, from June

Strenzel pinx. Weber Sculps.

1780 onwards, making a yearly issue (*année*) or volume of 48 plates with an engraved table of contents for each year except the 13th; plates 601 and 602 are lacking from most copies and apparently were not published until 1788, but careful reproductions of them, together with a table of contents for the 13th volume, were issued in 1840 by F. V. Raspail in his *Reproduction des 601 et 602me Planches qui manquent habituellement aux Champignons de Bulliard.* Plate 585 (*Agaric* d'Aulne) is numbered 581. It was the intention of the author that the *Atlas* when complete should serve to illustrate five books (*divisions*), of which, owing to his sudden death in 1793, only two were issued: *Histoire des Plantes vénéneuses et suspectes de la France* (Paris, 1784) and *Histoire des Champignons de la France* (Paris, 1791); no text was issued for *Plantes médicinales.* The *Herbier* is probably the first botanical work completely colour-printed without retouching by hand; see Bulliard, *Hist. Champignons* 336, footnote.

Publication seems to have been more or less as follows:

Vol. (année) 1, nos. 1–12, pls. 1–48 in June 1780–1781; vol. 2, nos. 13–24, pls. 49–96 in 1781–82; vol. 3, nos. 25–32, pls. 97–128 in 1782—Feb. 1783, nos. 32–36, pls. 129–144 in 1783; vol. 4, nos. 37–48, pls. 145–192 in 1783–84; vol. 5, nos. 49–60, pls. 193–240 in 1785; vol. 6, nos. 61–64, pls. 241–256 in 1785—Feb. 1786, nos. 65–72, pls. 256–288 in 1786; vol. 7, nos. 73–77, pls. 289–308 in March?—July 1787, nos. 78–84, pls. 309–336 in 1787–88; vol. 8, nos. 85–96, pls. 337–384 in 1788; vol. 9, nos. 97–108. pls. 385–432 in 1789; vol. 10, nos. 109–120, pls. 433–480 in 1790; vol. 11, nos. 121–132, pls. 481–528, in 1791; vol. 12, nos. 133–144, pls. 529–576 in 1792; vol. 13, nos. 145–150, pls. 577–600 in 1793; no. 151, pls. 601–602 uncertain, perhaps 1793 or 1798, certainly in 1840; see E. J. Gilbert, 'Un esprit, une œuvre; Bulliard, Jean Baptiste François, dit Pierre (1752–1793)', *Bull. Soc. Mycol. France* 68: 1–131, pls. 1–6 (1952).

BURBIDGE, FREDERICK WILLIAM THOMAS & BAKER, JOHN GILBERT

THE NARCISSUS, ITS HISTORY AND CULTURE . . . TO WHICH IS ADDED A SCIENTIFIC REVIEW OF THE ENTIRE GENUS

8vo (24 cm × 16½ cm). London 1875.

48 hand-coloured lithographed plates by F. W. Burbidge.

BUREAU, EDOUARD

MONOGRAPHIE DES BIGNONIACÉES . . .

4to with Atlas (27½ cm × 20 cm). Paris 1864.

30 uncoloured plates (pls. 1–31, pl. 22 missing) by A. Faguet, engraved by Picart, Annedouche, Debray, Lebrun.

BURGESS, HENRY W.

EIDODENDRON. VIEWS OF THE GENERAL CHARACTER AND APPEARANCE OF TREES . . .

Folio (57½ cm × 41 cm). London 1827 [–31].

54 uncoloured lithographed plates by H. W. Burgess and a portrait by W. C. Ross.

Some plates are dated as follows:—1–10, 12 (1827), 15–18 (1828), 21, 22, 24, 25 (1829), 34–36 (1830), 38 (1831). The rest bear no dates.

BURMANN, JOHANNES

THESAURUS ZEYLANICUS, EXHIBENS PLANTAS IN INSULA ZEYLANA NASCENTES . . . OMNIA ICONIBUS ILLUSTRATA . . .

4to (26½ cm × 22 cm). Amstelaedami (Amsterdam) 1737.

111 (pls. 1–110, 18*) unsigned engraved plates. Title-page drawn and engraved by A. van der Laan, portrait by J. M. Quinkhard engraved by J. Houbraken.

BURMANN, JOHANNES

RARIORUM AFRICANARUM PLANTARUM AD VIVUM DELINEATARUM ICONIBUS ET DESCRIPTIONIBUS ILLUSTRATARUM, DECAS I–X.

4to (26 cm × 21 cm). Amstelaedami (Amsterdam) 1738–39.

100 unsigned, uncoloured engravings by Hendrik Claudius, title-page drawn and engraved by J. C. Philips

BURMANN, NICOLAAS LAURENS

FLORA INDICA: CUI ACCEDIT SERIES ZOOPHYTORUM INDICORUM, NEC NON PRODROMUS FLORAE CAPENSIS.

4to (26 cm × 21 cm). Lugduni Batavorum (Leyden) & Amstelaedami (Amsterdam) 1768.

67 engraved plates by A. van der Laan.

BURNETT, M. A.

PLANTAE UTILIORES, OR ILLUSTRATIONS OF USEFUL PLANTS EMPLOYED IN THE ARTS AND MEDICINE.

4 vols 4to (27 cm × 21 cm). London [1839–] 1842–50.

260 hand-coloured, unnumbered and unsigned plates by Miss M. A. Burnett. Part of text by Gilbert Thomas Burnett, brother of Miss Burnett.

PLATE 30

In the early nineteenth century,
the look of flower prints changed as new printmaking media were adopted.
This amaryllis, a line etching colored by hand with opaque watercolor,
looks like German prints of almost a century earlier.
Painted by Strenzel, it adorns Leopold Trattinick's Thesaurus botanicus
(Vienna 1805–19).

BURY, MRS EDWARD

A SELECTION OF HEXANDRIAN PLANTS BELONGING TO THE NATURAL ORDERS AMARYLLIDAE AND LILIACEAE . . .
10 Parts. Folio (67 cm × 50 cm). London 1831–34.
51 aquatint plates drawn by Mrs Bury, engraved and partly colour-printed, partly hand-coloured by Robert Havell.
Discussed in text by Wilfrid Blunt.

BUTE, JOHN STUART, 3RD EARL OF

BOTANICAL TABLES CONTAINING THE DIFFERENT FAMILIES OF BRITISH PLANTS.
9 vols 4to (28½ cm × 23 cm). London [1785].
566 hand-coloured engraved plates by John Miller. Only 12 copies issued for private distribution; one copy is in Lindley Library of the Royal Horticultural Society, and another at the British Museum (Natural History). The cost of this imposing, but otherwise disappointing, work is said to have been over £12,000. See J. Britten in *J. Bot, London* 5484–87 (1916).

BUXBAUM, JOHANN CHRISTIAN

PLANTARUM MINUS COGNITARUM CENTURIA I (–V), COMPLECTENS PLANTAS CIRCA BYZANTIUM ET IN ORIENTE OBSERVATAS.
4to (28½ cm × 21½ cm). Petropoli (St. Petersburg) 1728–40.
326 unsigned engraved plates: part 1 pls. 1–65; part 2 pls. 1–50; part 3 pls. 1–74; part 4 39 pls. numbered from 1–66; part 5 44 pls, tabs numbered 1–71 and Appen no. 1. Plates in parts I–III, hand-coloured. Appendix contains 43 engraved figures (Nos. 2–44).

CANDOLLE, ALPHONSE LOUIS PYRAMUS DE

MONOGRAPHIE DES CAMPANULÉES.
4to (25 cm × 20½ cm). Paris 1830.
20 uncoloured plates by A. de Candolle, Heyland, engraved by Anspach, Heyland and Millenet.

CANDOLLE, AUGUSTIN PYRAMUS DE

PLANTARUM SUCCULENTARUM HISTORIA OU HISTOIRE NATURELLE DES PLANTES GRASSES.
2 vols Folio (53½ cm × 34 cm) and 4to (33 cm × 24 cm). Paris [1798–1829].
182 colour-printed engravings by Redouté.
Most copies contain plates numbered 1–159, published 1798 or 1799 to 1804 or 1805, but 23 extra plates listed by W. T. Stearn in Sitwell & Madol, *Album de Redouté* 16 (1954) were published in 1829.

CANDOLLE, AUGUSTUS PYRAMUS DE

ASTRAGALOGIA; NEMPE ASTRAGALI, BISERRULAE ET OXYTROPIDIS, NEC NON PHACAE, COLUTAE ET LESSERTIAE, HISTORIA ICONIBUS ILLUSTRATA.
Folio (36 cm × 25 cm). Paris Anno XI. 1802.
50 uncoloured plates by Redouté, engraved by Berthauld, Goulet, Guyard, l'Epine, Massard, Milsan, Plée, Tardieu.

CANDOLLE, AUGUSTUS PYRAMUS DE

STROPHANTHUS, NOVUM GENUS EX APOCINEARUM FAMILIA, DESCRIPTUM ET ICONIBUS ILLUSTRATUM.
Folio (50 cm × 38 cm). Paris Anno XII. 1804.
5 plates by Redouté, engraved by Plée.

CANDOLLE, AUGUSTIN PYRAMUS DE

ICONES PLANTARUM GALLIAE RARIORUM NEMPE INCERTARUM AUT NONDUM DELINEATARUM.
1 part 4to (34½ cm × 25 cm). Paris 1808.
50 uncoloured plates by Turpin, engraved by Plée and V. Plée fils.

CANDOLLE, AUGUSTIN PYRAMUS DE

PLANTES RARES DU JARDIN DE GENÈVE
4 parts 4to (36 cm × 27½ cm). Geneva, 1825–29.
24 partly hand-coloured plates, some engraved, some lithographed, drawn by Heyland (one of them drawn by Melle. C. Chuit), engraved by Millenet, Anspach, Bouvier, Bovet and Heyland.

CANDOLLE, AUGUSTIN PYRAMUS DE

COLLECTION DES MÉMOIRES (1–10) POUR SERVIR À L'HISTOIRE DU RÈGNE VÉGÉTAL.
4to (25½ cm × 20½ cm). Paris 1828–38.
99 uncoloured plates by Heyland, engraved by Plée, N. L. Rousseau and Bessin.

CATESBY, MARK

THE NATURAL HISTORY OF CAROLINA, FLORIDA, AND THE BAHAMA ISLANDS: CONTAINING THE FIGURES OF BIRDS, BEASTS, FISHES, SERPENTS, INSECTS AND PLANTS: PARTICULARLY THE FOREST TREES, SHRUBS, AND OTHER PLANTS, NOT HITHERTO DESCRIBED, OR VERY INCORRECTLY FIGURED BY AUTHORS. TOGETHER WITH THEIR DESCRIPTIONS IN ENGLISH AND FRENCH TO WHICH ARE ADDED OBSERVATIONS ON THE AIR, SOIL, AND WATERS: WITH REMARKS UPON AGRICULTURE, GRAIN, PULSE, ROOTS, ETC. TO THE WHOLE IS PREFIXED A NEW AND CORRECT MAP OF THE

PLATE 31
These beautiful anemones were painted by Clara Maria Pope to illustrate Samuel Curtis's The Beauties of Flora *(Gamston, England 1806–20). Its ten aquatint etchings show familiar flowers against landscape backgrounds* à la *Thornton, but with a gentler touch.*

COUNTRIES TREATED OF.
2 vols Large folio (approx. 51 cm × 35 cm). London 1731–1743.
Appendix 1747.
220 hand-coloured engravings of plants, birds, fishes etc., by Catesby.
First published in 11 parts: Vol. 1, pls. 1–60 in 1730, pls. 61–80 in 1731, pls. 81–100 in 1732; vol. 2, pls. 1–40 in 1734, pls. 41–60 in 1736, pls. 61–80 in 1738, pls. 81–100 in 1743; *Appendix*, pls. 1–20 (pl. 19 of fishes numbered as 20) in 1747. New edition, revised by George Edwards 1748–56, and a further edition 1771 both have same number of plates.
Catesby's own engravings are amusing, but inferior. W.B.

CATESBY, MARK
HORTUS BRITANNO-AMERICANUS, OR A CURIOUS COLLECTION OF TREES AND SHRUBS, THE PRODUCE OF THE BRITISH COLONIES IN NORTH AMERICA, ADAPTED TO THE SOIL AND CLIMATE OF ENGLAND.
4to (36 cm × 21½ cm). London 1763.
17 coloured engraved plates, each with 4 figures, except first and last plates which have one figure each, by M. Catesby adapted from those in his *Nat. Hist. Carolina*. This may have been first issued in 1737.
Ed. 2. HORTUS EUROPAE AMERICANUS, OR A COLLECTION OF 85 CURIOUS TREES AND SHRUBS . . .
4to (36 cm × 29 cm). London 1767.
17 coloured engraved plates as above.

CAVANILLES, ANTONIO JOSÉ
ICONES ET DESCRIPTIONES PLANTARUM QUAE AUT SPONTE IN HISPANIA CRESCUNT AUT IN HORTIS HOSPITANTUR.
6 vols. Folio (33½ cm × 22 cm). Matriti (Madrid) 1791–1801.
601 plates (1–600, 500 bis) by Cavanilles, engraved by Alex Blanco, J. Fonseca, M. Gamborino, T. and V. Lopez Enguidanos and Sellier.
Stiff, mechanical engravings, but of great botanical importance. W.B.

CHAMBERET, JEAN BAPT. JOS. CÉSAR TYRBAS DE. See CHAUMETON, F. 1815.

CHANDLER, ALFRED. See BOOTH, W. B. 1831.

CHANDLER, ALFRED & BUCKINGHAM, EDWARD BOURNE
CAMELLIA BRITANNICA
4to (31 cm × 24½ cm). London 1825.
8 hand-coloured lithographed plates by A. Chandler.

CHAUMETON, FRANÇOIS PIERRE; CHAMBERET, JEAN BAPT. JOS. CÉSAR TYRBAS DE & POIRET, JEAN LOUIS MARIE
FLORE MÉDICALE (FLORE DU DICTIONNAIRE DES SCIENCES MÉDICALES).
7 vols 8vo (21½ cm × 13 cm). Paris 1815–20.
426 hand-coloured plates (lithographs in vols. 1–6, pls. 1–349, 33 bis, 54 bis, 83 bis, 112 bis, 120 bis, 123 bis, 129 bis, 148 bis, 231 bis, 307 bis, 333 bis; mostly engravings in Vol. 7, pls. 1–56, 2 bis, 4 bis, suite de 4 bis, 36 bis, 43 bis, 44 bis, 48 bis, 56 bis, and 2 unnumbered plates) by Mme. E. Panckoucke and P.J.F. Turpin, lithographed and engraved by Lambert Jn., Bouteleu, Mme. Benoit, Bignant, Melle. Coignet, Melle. Cornu, Dien, Dubois, Forget, Gautier, Giraud, Goulot, Guyard, Melle. Louvier, Massard, Phelipeau, Plée, Poulet, Rebel and Victor.
Paris, 1814–20.
Also issued in 4to and folio, these agreeing in pagination but differing from the 8vo.

CHAZAL, ANTOINE
FLORE PITTORESQUE OU RECUEIL DE FRUITS ET DE FLEURS.
Folio Paris 1825.
70 plates printed in colour and finished by hand.
Chazal was a pupil of Van Spaëndonck and these plates are very good indeed. H.B.

CIRILLO, DOMENICO
PLANTARUM RARIORUM REGNI NEAPOLITANI FASCICULI I [–II].
Folio (36 cm × 24 cm). Neapoli (Naples) 1788–92.
24 hand-coloured plates by D. Cirillo, engraved by J. Brun and Ang. de Clener.

COINCY, AUGUSTE DE
ECLOGA I (–V) PLANTARUM HISPANICARUM SEU ICONES SPECIERUM NOVARUM VEL MINUS COGNITARUM PER HISPANIAS NUPERRIME DETECTARUM.
5 parts Folio (36 cm × 27½ cm). Paris 1893–1901.
59 uncoloured lithographed plates by B. Herincq.

PLATE 32
These noble carnations adorn the printed garden of nurseryman Samuel Curtis.
The Beauties of Flora *(Gamston, England 1806–20) is illustrated with aquatint etchings, a printmaking medium which describes areas of shade and not lines, thus creating the sky and background of this print.*

COMMELIN, CASPAR

PRAELUDIA BOTANICA AD PUBLICAS PLANTARUM EXOTICARUM DEMONSTRATIONES, DICTA IN HORTO MEDICO, 1701 & 1702 . . . ICONES ET DESCRIPTIONES.

4to (23½ cm × 18 cm). Lugduni Bat. (Leyden). 1703.
34 uncoloured engraved plates by P. Sluyter. Another edition. 1715.

COMMELIN, JAN, & COMMELIN, CASPAR

HORTI MEDICI AMSTELODAMENSIS RARIORUM . . . PLANTARUM, MAGNO STUDIO AC LABORE, SUMPTIBUS CIVITATIS AMSTELODAMENSIS, LONGA ANNORUM SERIE COLLECTARUM, DESCRIPTIO ET ICONES AD VIVUM AERI INCISAE.

2 vols Folio (40 cm × 26 cm). Amstelodami (Amsterdam) 1697 (Vol. 1), 1701 (Vol. 2).
213 unsigned uncoloured (rarely hand-coloured) engraved plates (including 2 pictorial title-pages and 5 plates representing coats of arms).
The first volume by Jan Commelin is a posthumous work edited by F. Ruysch and F. Kiggelar and has the words 'rariorum tam orientalis quam occidentalis Indiae, aliarumque peregrinarum plantarum' in its title. The second volume (*Pars altera. Tweede Deel*) by his nephew Caspar Commelin has the words 'rariorum tam Africanarum quam utriusque Indiae aliarumque peregrinarum plantarum' in its title.

COOKSON, MRS JAMES

FLOWERS DRAWN AND PAINTED AFTER NATURE IN INDIA

Large Folio (58 cm × 41 cm). London c. 1834–5.
31 hand-coloured lithographs. H.B.

COSSON, ERNEST SAINT-CHARLES & BARRATTE, J. FR. GUSTAVE

ILLUSTRATIONES FLORAE ATLANTICAE SEU ICONES PLANTARUM NOVARUM, RARIORUM VEL MINUS COGNITARUM IN ALGERIA NEC NON IN REGNO TUNETANO ET IMPERIO MAROCCANO NASCENTIUM.

2 vols 4to (35 cm × 27 cm). Paris 1882–97.
177 lithographed plates (pls. 1–175, 57 bis, 64 bis) by Ch. Cuisin and A. Riocreux.
For dates of publication see W. T. Stearn in *J. Soc. Bibl. Nat.* 1:149 (1938); Merrill in *J. Arnold Arb.* 22:455 (1941).
Fine botanical drawings. W.B.

CRADOCK, HON. MRS (HARRIET GROVE)

THE CALENDAR OF NATURE OR THE SEASONS OF ENGLAND EDITED WITH A PREFACE BY THE RT. HON. LORD JOHN RUSSELL

4 parts Folio (35 cm × 25 cm). London n.d. (1852).
Coloured title-page and 24 hand-coloured lithographs. H.B.

CURTIS, H.

THE BEAUTIES OF THE ROSE

2 vols. (Large paper edition). Folio (40 cm × 26 cm). (Ordinary edition). 4to (27 cm × 20 cm). Bristol 1850–1853.
38 hand-coloured lithographs. H.B.

CURTIS, SAMUEL

THE BEAUTIES OF FLORA, BEING A SELECTION OF FLOWERS PAINTED FROM NATURE . . .

Folio (69 cm × 54 cm). Gamston, Notts [1806–] 1820.
10 unsigned plates, partly hand-coloured, partly colour-printed, by Clara Maria Pope and T. Baxter (who also drew the title page), engraved by J. Hopwood, Weddell and F. C. Lewis (who aquatinted the background). This was first issued with the title *The Beauties of Flora, being a Selection of flowers painted from Nature by Thomas Baxter . . .*
Published by S. Curtis, Florist, Walworth, 1806.
Discussed by Wilfrid Blunt in text.

CURTIS, SAMUEL

A MONOGRAPH OF THE GENUS CAMELLIA.

Folio (70 cm × 57 cm). London 1819.
5 aquatint plates by Clara Maria Pope, engraved by Weddell.
One of the earliest and probably the finest of all the great Camellia books. In addition to the magnificent plates the directions in the text for cultivation and propagation are very sound. Discussed by Wilfred Blunt in text.

CURTIS, WILLIAM

FLORA LONDINENSIS, OR PLATES AND DESCRIPTIONS OF SUCH PLANTS AS GROW WILD IN THE ENVIRONS OF LONDON . . .

2 Vols Folio (47 cm × 28 cm). London [1775–] 1777–98.
434 hand-coloured plates by Sydenham Edwards, Will Kilburn and J. Sowerby, engraved by W. Darton and Co., W. Kilburn, F. Sansom, J. Sowerby and J. Swan. For references to papers on dates of publication, see M. J. van Steenis-Kruseman and W. T. Stearn in *Flora Malesiana* I.4; clxxvii (1954).
A splendid work, discussed in text by Wilfrid Blunt.

CURTIS, WILLIAM

FLORA LONDINENSIS CONTAINING A HISTORY OF THE PLANTS INDIGENOUS TO GREAT BRITAIN, ILLUSTRATED BY FIGURES OF THE NATURAL SIZE . . . A NEW EDITION ENLARGED BY GEORGE GRAVES AND WILLIAM JACKSON HOOKER.

5 vols Folio (49 cm × 30 cm). London 1817–28.
647 hand-coloured plates (illustrating 659 plants) by Sydenham Edwards, Will Kilburn, J. Sowerby, K. Greville, George Graves and William T. Hooker, engraved by W. Darton and Co., W. Kilburn, F. Sansom,

J. Sowerby, J. Swan and W. C. Edwards or lithographed by Pellitier.
Another edition of the above, enlarged by G. Graves and W. J. Hooker, 1817–28.

CURTIS, WILLIAM
LECTURES ON BOTANY AS DELIVERED IN THE BOTANIC GARDEN AT LAMBETH. INCLUDING A SKETCH OF THE LIFE AND WRITINGS OF THE LATE MR WILLIAM CURTIS BY DR THORNTON
3 vols 8vo (23 cm × 13 cm). London [1803]–1805.
119 hand-coloured engravings.
2nd edition 1807.

DECAISNE, JOSEPH, & LE MAOUT, EMMANUEL.
LE JARDIN FRUITIER DU MUSÉUM, OU ICONOGRAPHIE DE TOUTES LES ESPÈCES ET VARIÉTÉS, D'ARBRES FRUITIERS CULTIVÉS DANS CET ÉTABLISSEMENT . . .
9 vols Folio (30½ cm × 22 cm). Paris 1858–75.
508 colour-printed engraved or lithographed plates by A. Riocreux engraved by Melle. Taillant and P. Picart, chromolithographed by G. Severeyns.
A fine work. Alfred Riocreux (1820–1912) was the French equivalent of W. H. Fitch. W.B.

DE CANDOLLE, A. P. & A.L.P. See CANDOLLE, A. P. DE

DELESSERT, BENJAMIN
ICONES SELECTAE PLANTARUM QUAS IN SYSTEMATE UNIVERSALI, EX HERBARIIS PARISIENSIBUS, PRAESERTIM EX LESSERTIANO, DESCRIPSIT AUG. PYR. DE CANDOLLE, EX ARCHETYPIS SPECIMINIBUS A P. J. F. TURPIN DELINEATAE.
5 vols Folio (35 cm × 26 cm). Paris 1820–46.
501 plates by Borromée, J. Decaisne (analyses), E. Delile, Heyland, Node-Véran, Riocreux and Turpin, engraved by 21 engravers. Each volume has 100 plates except vol. 5 which has 101, there being a 93A & 93B.

DENISSE, ETIENNE
FLORE D'AMERIQUE. DESSINÉE D'APRÈS NATURE SUR LES LIEUX . . .
12 parts Folio Paris (1843–46).
72 hand-coloured lithographs. H.B.

DE PUYDT, PAUL EMILE
LES ORCHIDÉES, HISTOIRE ICONOGRAPHIQUE.
Royal 8vo. Paris 1880.
50 chromolithographs. H.B.

DESCOURTILZ, MICHEL ETIENNE
FLORE PITTORESQUE ET MÉDICALE DES ANTILLES, OU HISTOIRE NATURELLE DES PLANTES USUELLES DES COLONIES FRANÇAISES, ANGLAISES, ESPAGNOLES ET PORTUGAISES, PEINTE D'APRÈS LES DESSEINS FAITS SUR LES LIEUX.
8 vols 8vo (21 cm × 12½ cm). Paris 1821–29.
600 partly colour-printed, partly hand-coloured plates by J. Théodore Descourtilz, engraved by Gabriel, Perée, Bessin, Prieur. Copies also exist with plates and a greater part of the text on larger paper.
2nd Ed. Paris. 1827–33.

DE SÈVE, JACQUES EUSTACE
RECUEIL DE VINGTQUATRE PLANTES ET FLEURS; D'APRÈS LES DESSINS DE M. SEVE.
Folio (about 45 cm × 30 cm). Paris, after 1772.
24 numbered hand-coloured engraved plates.

DESFONTAINES, RENÉ LOUICHE
FLORA ATLANTICA, SIVE HISTORIA PLANTARUM, QUAE IN ATLANTE, AGRO TUNETANO ET ALGERIENSI CRESCUNT.
2 vols 4to (29 cm × 21 cm). Paris 1798–99.
263 plates (1–261, 76 bis, and a plate called 'pag. 254' inserted between pls. 254 and 255) by Maréchal, P. J. Redouté, Fossier, H. J. Redouté and Malœuvre. Engraved by Guyard and Malœuvre. For dates of publication see W. T. Stearn in *J. Soc. Bibl. Nat. Hist.* I. 147 (1938). Three versions of the title page exist.

DESFONTAINES, RENÉ LOUICHE
CHOIX DES PLANTES DU COROLLAIRE DES INSTITUTS DE TOURNEFORT, PUBLIÉES D'APRÈS SON HERBIER, ET GRAVÉES SUR LES DESSINS ORIGINAUX D'AUBRIET.
4to (29½ cm × 22½ cm). Paris 1808.
70 plates (1–69, and one unnumbered) by Aubriet, engraved by Lambert.

DICKSON, R. W. See MACDONALD, ALEXANDER

DIETRICH, ALBERT GOTTFRIED
FLORA REGNI BORUSSICI. FLORA DES KÖNIGREICHS PREUSSEN ODER ABBILDUNG UND BESCHREIBUNG DER IN PREUSSEN WILDWACHSENDEN PFLANZEN.
12 vols 8vo (26 cm × 16 cm). Berlin 1833–43.
864 unsigned hand-coloured lithographed plates by A. G. Dietrich.

DIETRICH, DAVID (NATHANIEL) FRIEDRICH
FLORA UNIVERSALIS IN COLORIRTEN ABBILDUNGEN.
15 vols Folio (40 cm × 26 cm). Jena 1831–54 (–55).
4760 plates by D.N.F. Dietrich, engraved by D.N.F. Dietrich, F. Kirchner, W. Müller.
Neue Folge, welche grosstentheils neu entdeckte noch nicht abgebildete Pflanzen enthält. 9 parts. Leipzig 1849 [–55]. 90 plates as above. Neue Serie, 1861. 2 parts with 20 plates.

Abricot-pêche.

104.

Turpin Pinx. De l'Imprimerie de Langlois. Bouquet Sculp.

DIETRICH, DAVID (NATHANIEL) FRIEDRICH

DEUTSCHLANDS ÖKONOMISCHE FLORA ODER BESCHREIBUNG UND ABBILDUNG ALLER FÜR LAND- UND HAUSWIRTHE WICHTIGEN PFLANZEN.
3 vols 8vo (17½ cm × 10½ cm). Jena 1841–44.
147 unsigned engraved plates (Nos. 1, 1–146) all hand-coloured except No. 1, engraved by A. Hanemann, C. Hensgen, F. Kirchner, and Mädel II.

DIETRICH, DAVID (NATHANIEL) FRIEDRICH

FORST-FLORA. BESCHREIBUNG UND ABBILDUNG ALLER FÜR DEN FORSTMANN WICHTIGEREN WILDWACHSENDEN BÄUME UND STRÄUCHER . . .
Ed. 6 by Felix von Thümen.
2 vols 4to (30 cm × 21 cm). Dresden [1885–86] 1887.
300 unsigned hand-coloured engraved plates, engraved by A. Hanemann and F. Kirchner.

DILLENIUS, JOHANN JAKOB

HORTUS ELTHAMENSIS, SEU PLANTARUM RARIORUM QUAS IN HORTO SUO ELTHAMI IN CANTIO COLUIT JACOBUS SHERARD DELINEATIONES ET DESCRIPTIONES.
2 vols Folio (47 cm × 29 cm). London 1732.
325 engraved plates by J. J. Dillenius. (Pls. 1–324 and one inserted between plates 166 and 167 facing p. 206.)
Usually uncoloured but a few copies hand-coloured.
A most lavishly-illustrated catalogue of the plants in James Sherard's garden at Eltham. Interesting as a record from such an early date and notable for the number of illustrations of succulent plants.
HORTI ELTHAMENSIS PLANTARUM RARIORUM ICONES ET NOMINA.
2 vols Folio (41 cm × 25 cm). Lugduni Bat. (Leyden) 1774.
325 engraved plates by J. J. Dillenius.
This Leyden edition is without text.

DODART, DENIS

RECUEIL DES PLANTES.
3 vols. Folio (63 cm × 46½ cm). Paris (Imprimerie Royale) 1701.
Another edition, 1788.
319 engraved plates, arranged alphabetically, by Nicolas Robert.
This great work was started in 1675 and an elaborate 'project' was issued then under the title *Mémoires pour servir à l'histoire des Plantes. Dressez par M. Dodart.*
Discussed in text by Wilfrid Blunt, also in his book on Botanical Illustration, pp. 111, 112.

DRAKE DEL CASTILLO, EMMANUEL

ILLUSTRATIONS FLORAE INSULARUM MARIS PACIFICI.
4to (31 cm × 23½ cm). Parisiis (Paris) 1886 [–92].
50 plates drawn and lithographed by d'Apréval.
For dates of publication see W. T. Stearn in *J. Soc. Bibl. Nat. Hist.* 1. 202 (1939).

DREVES, JOHANN FRIEDRICH PETER & HAYNE, FRIEDRICH GOTTLOB

BOTANISCHES BILDERBUCH.
5 vols 4to. (24 cm × 18 cm). Leipzig 1794–1801.
152 hand-coloured engravings by Cabieux and Guimpel.
Also a French edition, Leipzig 1802, with 152 hand-coloured plates, with Friedrich Gottlob Hayne's name also appearing on the title-page. H.B.

DUFOUR, AUGUSTINE

L'ART DE PEINDRE LES FLEURS À L'ACQUARELLE . . . ORNÉ D'UN CHOIX DES PLUS BELLES FLEURS, GRAVÉES D'APRÈS LES DESSINS DE MLLE AUGUSTINE DUFOUR, ÉLÈVE DE REDOUTÉ.
4to (37½ cm × 24 cm). Paris 1834.
36 coloured plates by Augustine Dufour, engraved by Mlle Perrot.

DUHAMEL DU MONCEAU, HENRI LOUIS

TRAITÉ DES ARBRES ET ARBUSTES QUI SE CULTIVENT EN FRANCE EN PLEINE TERRE.
2 vols 4to. Paris 1755.
250 uncoloured woodcut plates and numerous text figures.
New edition 1785.

DUHAMEL DU MONCEAU, HENRI LOUIS

TRAITÉ DES ARBRES ET ARBUSTES QUE L'ON CULTIVE EN PLEINE TERRE EN FRANCE, SECONDE ÉDITION CONSIDÉRABLEMENT AUGMENTÉE [NOUVEAU DUHAMEL].
7 vols Folio (42 cm × 26 cm). Paris [1800] 1804–19.
498 colour-printed plates by P. J. Redouté and P. Bessa, engraved by 54 engravers.
Published in 86 parts. For dates of publication and list of contributing authors see W. T. Stearn in Sitwell and Madol, *Album de Redouté* 16 (1954). Virtually an independent work from the preceding with enlarged text and new plates.
Reissued 1852 or 1853.
Colour-printed plates not retouched by hand and therefore lacking quality. W.B.

PLATE 33
Henri Duhamel du Monceau's Traité des arbres fruitiers *(Paris 1807–35) catalogues the horticulture of France with 422 jewel-like prints like "Abricot-pêche." Pierre Turpin's watercolor was etched in stipple by Bouquet, and printed in five colors* à la poupée *at the press of Langlois. The 1768 first edition of this work was illustrated with line engravings printed in black.*

Fico Fetifero, o Fico dell'Osso

Domenico Del Pino disegnò in Finale 1826.

Giuseppe Pera inc. in Firenze

DUHAMEL DU MONCEAU, HENRI LOUIS
TRAITÉ DES ARBRES FRUITIERS.
2 vols 4to (34 cm × 25 cm). Paris 1768.
About 180 uncoloured engravings. Plates by Aubriet and other artists.
New edition, by A. Poiteau and P.J.F. Turpin.
Folio 6 vols. Paris 1807–1835.
422 plates printed in colour, and retouched by hand, from paintings by Turpin and Poiteau with numerous engravings.
The plates republished 1846 by Poiteau alone, under the title *Pomologie Française*.

DUJARDIN-BEAUMETZ & ÉGASSE, E.
LES PLANTES MÉDICINALES INDIGÈNES ET EXOTIQUES . . .
4to (28 cm × 19 cm). Paris 1889.
40 chromolithographed plates by A. Lefèvre, chromolithographed by Monrocq, and 1034 engraved text-figs. by Cordier, Hugon, E. Oberlin, Thiebault, Storck, Marchand, Michelet and others.

DUPPA, RICHARD
THE CLASSES AND ORDERS OF THE LINNAEAN SYSTEM OF BOTANY.
3 vols 8vo. London 1816.
240 hand-coloured engravings. H.B.

DUPPA, RICHARD
ILLUSTRATIONS OF THE LOTUS OF ANTIQUITY.
Folio. London 1813.
12 hand-coloured engravings, one from a drawing by P. J. Redouté. H.B.
Expanded edition, slightly different title, 1816.

EDWARDS, JOHN
THE BRITISH HERBAL, CONTAINING ONE HUNDRED PLATES OF THE MOST BEAUTIFUL AND SCARCE FLOWERS AND USEFUL MEDICINAL PLANTS . . . OF GREAT BRITAIN, ACCURATELY COLOURED FROM NATURE . . .
Folio (49 cm × 30 cm). London (1769–) 1770.
100 hand-coloured plates by J. Edwards, engraved by J. Edwards, Wm. Darling and J. Fougeron.
Also issued as *Edward's Herbal containing the most beautiful and scarce Flowers* (1770) and later re-issued as *A select Collection of one hundred Plates consisting of the most beautiful exotic and British Flowers* . . . (1775). Plates and text are the same in all three but the colouring varies in quality.
Decorative but stylised plates. W.B.

EDWARDS, JOHN
A COLLECTION OF FLOWERS DRAWN AFTER NATURE AND DISPOSED IN AN ORNAMENTAL AND PICTURESQUE MANNER.
Folio (52 cm × 36 cm). London 1783–95, and 1801.
Uncoloured engraved title page acting as plate 1, and 79 hand-coloured engraved plates by John Edwards.
No text.
Decorative but stylised plates. W.B.

EDWARDS, SYDENHAM
THE NEW FLORA BRITANNICA.
2 vols. 4to (30 cm × 23 cm). London, 1812.
61 hand-coloured engravings published first in Macdonald's *A Complete Dictionary of Practical Gardening* 1807 (Q. V.) and later used under the title of THE NEW BOTANIC GARDEN. H.B.

EDWARDS, SYDENHAM. See Periodicals.

EEDEN, A. C. VAN
ALBUM VAN EEDEN. FLORA OF HAARLEM. COLOURED PLATES OF DUTCH BULBS AND BULBOUS PLANTS.
Folio (36½ cm × 27½ cm). Haarlem 1872–81.
120 chromolithographed plates by Melle. Arentine H. Arendsen, lithographed by G. Severeyns.

EHRET, GEORG DIONYS
[PLANTAE ET PAPILIONES RARIORES DEPICTAE ET AERI INCISAE]
Folio (50 cm × 35 cm). London [1748–59].
15 hand-coloured engraved plates by G. D. Ehret. Plate 1 dated '1748', plate 15 '1759'.
There is no title-page; the title cited is the heading of plate 1; plate 15 has a page of text entitled 'The history and Analysis of the Parts of the Jessamine'.
Splendid Ehret plates. W.B.

EICHELBERG, JOHANN FRIEDRICH ANDREAS
NATURGETREUE ABBILDUNGEN UND AUSFÜHRLICHE BESCHREIBUNGEN ALLER IN- UND AUSLÄNDISCHER GEWÄCHSE, WELCHE DIE WICHTIGSTEN PRODUCTE FÜR HANDEL UND INDUSTRIE LIEFERN . . .
8vo (26½ cm × 16 cm). Zurich 1845.
72 hand-coloured engraved and lithographed plates by Eichelberg, engraved and lithographed by J. Kull.

PLATE 34
Giorgio Gallesio harnessed color-printed stipple etching for his Pomona italiana *(Pisa 1817–39), a book of around 170 plates of fruit. All artists and craftsmen involved were Italian, and yet this fine work is almost isolated in Italian publishing, much as Hoffmannsegg (Plates 39 and 40) is among German books.*

ELWES, HENRY JOHN
A MONOGRAPH OF THE GENUS LILIUM
Folio (54½ cm × 37½ cm). London [1877–] 1880.
48 hand-coloured plates drawn and lithographed by W. H. Fitch. 1 map, 1 photograph.
Supplement by Arthur Grove and A. D. Cotton.
7 parts. Folio (54½ cm × 37½ cm). London 1934–40.
30 hand-coloured plates by Lilian Snelling.
For dates of publication, see Woodcock & Stearn, *Lilies of the World* 397 (1950).
A work of the greatest interest and value to all Lily growers. The Supplement is considerably finer both in text and illustration than the original monograph. P.M.S.

ENDLICHER, STEPHAN LADISLAUS
ATAKTA BOTANIKA. NOVA GENERA ET SPECIES PLANTARUM DESCRIPTA ET ICONIBUS ILLUSTRATA.
Folio (47½ cm × 30½ cm). Vindobonae (Vienna) 1835 [–35].
35 uncoloured plates (1–40, of which pls. 10, 26, 28, 37, 38 did not appear) by Ferd. Bauer, Endlicher, E. Fenzl, F. X. Fieber and J. Zehner, engraved by M. Bauer, Berkowetz, C. Neunlist and Jos. Skala.
For dates of publication, see W. T. Stearn in *J. Arnold Arb*. 38: 426 (1947).

ENDLICHER, STEPHAN LADISLAUS
ICONOGRAPHIA GENERUM PLANTARUM
10 parts 4to (28 cm × 21½ cm). Vindobonae (Vienna) [1837–] 1838 [–41].
125 uncoloured plates by Ferd. Bauer, Ed. Fenzl, Aloys Putterlink, Jos. Zehner, engraved by Gebhardt.
For dates of publication see W. T. Stearn in *J. Arnold Arb*. 28:425 (1947).

ENDLICHER, STEPHAN. See HARTINGER, A. 1844, POEPPIG, E. F. 1835, SCHOTT, H. W. 1832.

EVERARD, ANNE
FLOWERS FROM NATURE WITH THE BOTANICAL NAME, CLASS AND NATURE
Small Folio. London 1835.
13 hand-coloured lithographs. H.B.

FISCHER, FRIEDRICH ERNST LUDWIG, & OTHERS
JARDIN DE SAINT PÉTERSBURG. SERTUM PETROPOLITANUM, SEU ICONES ET DESCRIPTIONES PLANTARUM QUAE IN HORTO BOTANICO IMPERIALI PETROPOLITANO FLORUERUNT
4 parts Folio (49 cm × 32 cm). St. Petersburg 1846–49.
39 plates, some lithographed, some engraved, by J. A. Satory, lithographed or engraved by J. A. Satory and G. Hesse; also 7 plans and 1 view of glasshouses.
Decas I (1846), by F.E.L. Fischer and C. A. Meyer; Decas II by C. A. Meyer (1852); Decas III and IV by E. Rege (1869).
Plates 1–20 hand-coloured, 16 engraved, 4 (pls. 12, 13, 18, 19) lithographed; pls. 21–39 lithographed and usually uncoloured (pl. 28 sometimes hand-coloured).

FITCH, W. H. See HOOKER, SIR W. J. 1847 & 1851.

FITTON, E. & S. M., CONVERSATIONS ON BOTANY. See ANONYMOUS 1817.

FITZGERALD, ROBERT DAVID
AUSTRALIAN ORCHIDS
2 vols Folio (48 cm × 33½ cm). Sydney (1875–) 1882 [–1894].
119 unnumbered plates (most of them hand-coloured) by R. D. Fitzgerald, lithographed by R. D. Fitzgerald and Arthur J. Stopps.
Although the title-page of Vol. 1 is dated '1884', James Norton's preface to Vol. 2 part 5 states that: 'the first part . . . was published by the Government Printer in July 1875, since which time the first volume consisting of seven numbers, comprising about ten plates each . . . has been completed, and up to March 1888 four parts of the second volume had been issued'. Fitzgerald died on 12 August 1892 and Vol. 2 part 5 was completed from his notes by Henry Dane in 1894. The contents of all the parts are listed on the back wrapper of Vol. 2 part 5.

FLINDERS, MATTHEW
A VOYAGE TO TERRA AUSTRALIS . . . IN THE YEARS 1801, 1802 AND 1803, IN H.M.S. THE INVESTIGATOR . . .
2 vols 4to (30 cm × 23 cm); Atlas, Folio (66 cm × 47 cm). London 1814.
9 uncoloured plates representing scenery in the text, and 28 in the Atlas, of which 10 are botanical, by Ferd. Bauer and Westall, engraved by Eliz. Byrne, Sanson and I. Pye.

FOURREAU, JULES. See JORDAN, A. 1866.

FRANCHET, ADRIEN
PLANTAE DELAVAYANAE. PLANTES DE CHINE RECUEILLIES AU YUN-NAN PAR L'ABBÉ DELAVAY . . .
8vo (24½ cm × 16 cm). Paris 1889–90.
40 lithographed plates (numbered 1–46, of which 2–3, 9–10, 27–28, 29–30, 32–33 and 34–35 represent one plate each) by A. d'Apréval.
An important work botanically and horticulturally since in it are described and named the new plants collected by the Abbé Delavay in Yunnan, many of which have proved excellent garden plants. P.M.S.

FRYER, ALFRED & BENNET, ARTHUR

THE POTAMOGETONS (PONDWEEDS) OF THE BRITISH ISLES.

4to (31 cm × 24 cm). London [1898–] 1915.

60 hand-coloured plates mostly by Robert Morgan (pls. 1–48, 56, 59, 60), the others by E. C. Knight, and Miss M. Smith, lithographed by R. Morgan, E. C. Knight and M. Smith.

Parts 1–6, pp. 1–36, pls. 1–24 in 1898; parts 7–9, pp. 37–56, pls. 25–36, in 1900; parts 10–12, pp. 57–76, pls. 37–49, in 1913; parts 13–15, pp. 77–94, i–xi, pls. 50–60, in 1915.

FURBER, ROBERT

TWELVE MONTHS OF FLOWERS

Folio (58½ cm × 47 cm). London 1730.

13 hand-coloured engraved plates (1 of which gives the list of subscribers and serves as title-page, the remaining twelve, each representing one month of the year, being flower-arrangements in vases, showing in all 400 flowers) by P. Casteels, engraved by H. Fletcher.

A rare but grossly over-rated work, notable as the first attempt to convert by means of engraving the many-bloomed Dutch flower-piece into a book illustration or an advertising broadsheet. That these engravings do little justice to Pieter Casteels as a flower-painter is evident from the set of 12 original paintings by Casteels (made from the drawings used by the engraver) formerly in the collection of Lord Methuen at Corsham Court and sold by Lady Oppenheimer at Sothebys in February 1950. Reduced copies of these engravings illustrate The Flower-Garden Display'd *(1732 and 1734) commonly listed under Furber's name but best attributed to Richard Bradley, q.v.*

The book was really a very elaborate nurseryman's catalogue. Many of the flowers allotted to the autumn and winter months do not, however, usually flower then, a license similar to that taken by many of the old Dutch painters in constructing their pictures.

FURBER, ROBERT. See [BRADLEY, R.] 1730.

FURBER, ROBERT

FLORA, OR A CURIOUS COLLECTION OF YE MOST BEAUTIFUL FLOWERS . . . IN . . . EACH MONTH OF THE YEAR . . .

Folio. London 1749 and 1770.

Title page and 12 coloured plates, engraved by Clark and Parr.

FURBER, ROBERT

THE TWELVE MONTHS OF FLOWERS

copied (in reverse) and re-engraved on a smaller scale, published by six different London print publishers in two different issues of the plates.

4to. London n.d.

Title-page and 12 plates, engraved by R. Sayer.

GALLESIO, GIORGIO

POMONA ITALIANA, OSSIA TRATTATO DEGLI ALBERI FRUTTIFERI.

2 vols. Folio (48½ cm × 32 cm). Pisa 1817 (–39).

Up to 170 (according to Arnold Arboretum Catalogue) colour-printed plates by D. del Pino, A. Serantoni and many others, engraved by several engravers. Copies vary in the number of plates, which are unnumbered and differently arranged in different copies.

GARCKE, FRIEDRICH AUGUST. See KLOTZSCH, J. F. 1862.

GARDENERS, SOCIETY OF

CATALOGUS PLANTARUM . . . A CATALOGUE OF TREES, SHRUBS, PLANTS AND FLOWERS . . . PROPAGATED . . . NEAR LONDON . . . BY A SOCIETY OF GARDENERS.

Folio (44 cm × 28½ cm). London 1730.

Frontispiece, 2 vignettes and 21 coloured engraved plates (with 57 figures) by Jacob v. Huysum, engraved by Henry Fletcher and E. Kirkall.

For a note on this work by J. Ardagh, see *J. Bot. (London)* 72: 111 (1934).

The text was probably written by Philip Miller who was secretary to the Society of Gardeners, about twenty members strong, which for several years (c. 1720–30) met monthly near London; at these meetings 'each Person of the Society brought all the several kinds of Plants, Flowers and Fruits in their various Seasons, which were there examined and compared by all the Persons present'.

GAUTIER-DAGOTY, JACQUES

COLLECTION DES PLANTES USUELLES, CURIEUSES ET ÉTRANGÈRES, GRAVEÉS ET IMPRIMÉES EN COULEURS . . .

Folio (about 41 cm × 25 cm). Paris 1767.

40 unnumbered plates, engraved by Gautier-Dagoty, printed in colour by the Le Blond three-colour process.

GAY, CLAUDE

HISTORIA FISICA Y POLITICA DE CHILE. BOTANICA.

8 vols 8vo (21½ cm × 13½ cm); Atlas. Folio (36 cm × 28½ cm). Santiago & Paris 1845–52 [–54].

103 uncoloured engraved plates (Phanerogams pls. 1–83, 32 duplicate number, 32 bis, 52 bis, 69 bis; Cryptogams pls. 1–16) by A. Riocreux, E. Desvaux, Oudart, J. Remy, Vauthier and Vianne, engraved by 16 engravers.

For dates of publication, see I. M. Johnston in *Darwiniana* 5: 154–165 (1941).

GEEL, P. C. VAN. See Periodicals.

GILIBERT, JEAN EMMANUEL

DÉMONSTRATIONS ÉLÉMENTAIRES DE BOTANIQUE (PAR LATOURETTE ET ROZIER).
Ed. 3, corrigée et augmentée par J. E. Gilibert.
3 vols 8vo. Lyon 1787.
13 engraved plates (Nos. 1–8 and 5 unnumbered) by various artists, engraved by Superchi.
Ed. 4. 4 vols 8vo. (Lyon 1796). 13 engraved plates by various artists.
Partie de figures 2 vols. 4to. 343 engravings in 5 series by Superchi.

GLEADALL, ELIZA EVE

THE BEAUTIES OF FLORA WITH BOTANIC AND POETIC ILLUSTRATIONS
2 vols 4to (30 cm × 25 cm). Published by Eliza Eve Gleadall at Heath Hall, Wakefield, 1834.
41 hand-coloured lithographs. H.B.

GOUAN, ANTOINE

ILLUSTRATIONES ET OBSERVATIONES BOTANICAE, AD SPECIERUM HISTORIAM FACIENTES, SEU RARIORUM PLANTARUM INDIGENARUM PYRENAICARUM, EXOTICARUM ADUMBRATIONES . . . CUM ICONIBUS EX NATURAE TYPO ET MAGNITUDINE NATURALI AB AUCTORE DELINEATIS.
Folio (42 cm × 25½ cm). Tiguri (Zurich) 1773.
28 engraved plates (Nos. i–xxvi, xxa, xxb) by Gouan.

GOURDON, JEAN & NAUDIN, P.

NOUVELLE ICONOGRAPHIE FOURRAGÈRE. HISTOIRE BOTANIQUE, ECONOMIQUE ET AGRICOLE DES PLANTES FOURRAGÈRES . . .
1 vol. and Atlas. 4to (28 cm × 20 cm). Paris [1865–] 1871.
125 hand-coloured engraved plates (Nos. 1–25, 46 bis, and Nos. 103, 104 represented by one double plate) by P. Naudin, engraved by A. Duménil and C. Pierre.

GRANDIDIER, ALFRED

HISTOIRE PHYSIQUE, NATURELLE ET POLITIQUE DE MADAGASCAR. HISTOIRE NATURELLE DES PLANTES.
5 vols (4 vols atlas; 1 vol. text) i.e.
Vols 29–30, 33–36 4to (31 cm × 24 cm). Paris 1886–1903.
553 lithographed plates (pls. 1–504, 11A, 16A, 23A, 24A–D, 30A–B, 44A–C, 49A, 79A–I, 170A, 226A–E, 226D bis, 230A–B, 245A, 248A–B, 250A, 252A, 254A–D, 262A–D, 280A–B, 281A, 282A–B, 283A, 292A–C, 293A, 294A–C, 308A, 336A, 345A, 347A, 412A, 415A–B, 428A, 437A–C, 442A–D) mostly by d'Apréval, but some (e.g. 248A, 249, 254A, 261, 262, 284) by A. Faguet and 79F–I from photographs.
Plate 501 never published. Plate 796 (labelled *Andansonia madagascariensis*, the analyses unnumbered) issued in 1889 was cancelled in 1893 by a revised plate 796 (labelled 1. *Andansonia madagascariensis*, 2. *A. Grandidieri*, 3. *A. digitata*).
The plates of this book were drawn under the supervision of H. E. Baillon up to his death in 1895 and he was responsible for the scientific names on the 370 plates then published. Many of them represented new species of which Baillon published descriptions in *Bull. Soc. Linn. Paris*. In 1896 E. Drake del Castillo assumed scientific responsibility for the work and later published a volume of text (vol. 33, tome 1, part 1, with 208 pages covering the families Ranunculaceae to Proteaceae) which is dated '1902' but was received at the British Museum (Nat. Hist.) on 24 Feb. 1903 and listed in *Nat. Nov.* 25 (6): 205 (March 1903).
Publication of the Atlas was as follows:
Tome 2 (Atlas 1), fasc. 14, pls. 1–50 in 1886 (listed in *Nat. Nov.* in Feb. 1887), fasc. 18, pls. 51–88, 11A, 16A, 23A, 24A–D, 30A, 30B, 44A–C in 1888 (listed in *Nat. Nov.* in March 1889), fasc. 20, pls. 89–130, 79A, 79B (*Adansonia madagascariensis*) in 1889 (listed in *Nat. Nov.* in Feb. 1890).
Tome 3 (Atlas 2), fasc. 22, pls. 131–160 in 1890 (listed in *Nat. Nov.* in March 1891), fasc. 27, pls. 161–204 in 1891 (listed in *Nat. Nov.* in March 1891), fasc. 30, pls. 205–241 in 1892, fasc. 34, pls. 242–260, 49A, 79B (*Adansonia madagascariensis*, *A. Grandidieri*, *A. digitata*), 79C–I, 226A–E, 230A, 230B in 1893 (listed in *Nat. Nov.* in May 1894).
Tome 4 (Atlas 3), fasc. 35, pls. 261–276, 284–290, 296–304, 321–324 in 1894 (listed in *Nat. Nov.* in Sept. 1894), fasc. 36, pls. 291–295, 292C, 293A, 294A–C, 305–308, 325–328, 339, 340 in 1895 (listed in *Nat. Nov.*' in April 1895), fasc, 38, pls. 262A–D, 309, 310, 312–314, 317–319, 329, 330, 332, 341–344, 354–356, 359, 361–365, 369, 370, in 1895 (listed in *Nat. Nov.* in Dec. 1895), fasc. 40, pls. 170A, 311, 315, 316, 332–338, 345–353, 345A, 347A, 357, 358, 360, 366–368, 371–374, 398 in 1896 (listed in *Nat. Nov.* in Sept. 1896), fasc. 5, pls. 248B, 277–283, 280A, 280B, 281A, 282A, 282B, 283A, 320, 331, 336A, 375–397, 399–411, in 1896 (listed in *Nat. Nov.* in Feb. 1897).
Tome 5 (Atlas 4) pls. 292A, 292B, 308A, 412–457, in 1897 (listed in *Nat. Nov.* in April 1898), fasc. 49, pls. 412A, 415A, 415B, 428A, 437A–C, 442A–D, 451A, 460–475, 477–479, 481 in 1899 (listed in *Nat. Nov.* in June–July 1900), fasc. 54, pls. 458, 459, 476, 480, 482–500, 502–504 in 1903 (listed in *Nat. Nov.* in Feb. 1904). Judging from the title-page Drake del Castillo is responsible for this volume.

GRAVES, GEORGE & MORRIES, JOHN DAVIE

HORTUS MEDICUS, OR FIGURES AND DESCRIPTIONS OF THE MORE IMPORTANT PLANTS USED IN MEDICINE . . .
8vo (27½ cm × 20½ cm). Edinburgh and London 1834.
44 hand-coloured engraved plates by G. Graves, engraved by W. H. Lizars.

GRAY, ASA

GENERA FLORAE AMERICAE BOREALI-ORIENTALIS ILLUSTRATA. THE GENERA OF THE PLANTS OF THE UNITED STATES, ILLUSTRATED BY FIGURES AND ANALYSES FROM NATURE BY ISAAC SPRAGUE, SUPERINTENDED AND WITH DESCRIPTIONS ETC.: BY ASA GRAY.

2 vols 8vo (23½ cm × 14 cm). Boston & New York 1848–49.

185 plates (pls. 1–186, pl. 40–41 being one plate) by Isaac Sprague, lithographed by J. Prestele.

GREEN, THOMAS

THE UNIVERSAL HERBAL . . .

2 vols 4to (27 cm × 21 cm). London 1816–20.

2 coloured frontispieces. 1 coloured title page (to Vol. 1). and 102 hand-coloured wood engravings by F. Dixon, G. Dobie, W. Swift and others. H.B.

GUILLEMIN, J. B. ANTOINE

ICONES LITHOGRAPHICAE PLANTARUM AUSTRALASIAE RARIORUM.

2 parts Folio (34½ cm × 25½ cm). Paris 1827.

20 unsigned lithographed plates.

GUILLEMIN, J. B. ANTOINE; PERROTTET, GEORGE SAMUEL & RICHARD, ACHILLE

FLORAE SENEGAMBIAE TENTAMEN, SEU HISTORIA PLANTARUM IN DIVERSIS SENEGAMBIAE REGIONIBUS A PEREGRINATORIBUS PERROTTET ET LEPRIEUR DETECTARUM.

Vol 1 (all published). Folio (33 cm × 25 cm). Paris 1830–33.

72 uncoloured plates by Decaisne and Vauthier, lithographed by Jouy and Vielle.

For dates of publication, see W. T. Stearn in *Flora Malesiana* I.4: clxxxv (1954). Although dated 1830–1833 on the title-page, publication seems to have begun in January 1831.

GUIMPEL, FRIEDRICH. See WILLDENOW, C. L. 1815.

GUIMPEL, FRIEDRICH. See HAYNE, F. G. 1825.

GUSSONE, GIOVANNI

FLORA SICULA SIVE DESCRIPTIONES ET ICONES PLANTARUM RARIORUM SICILIAE ULTERIORIS . . .

Folio (46½ cm × 32 cm). Neapoli (Naples) 1829.

6 hand-coloured engraved plates (Nos. i–v, xiv) by F. Paderni, engraved by C. Biondi. The last plate bears the number xiv.

HALLER, ALBRECHT VON

ICONES PLANTARUM HELVETIAE EX IPSIUS HISTORIA STIRPIUM HELVETICARUM DENUO RECUSAE . . . ADDITIS NOTIS EDITORIS (J.S. WYTTENBACH).

Folio (39½ cm × 24 cm). Bernae (Berne) 1795.

52 uncoloured plates on 46 sheets by C. J. Rollinus, engraved by C. F. Fritsch, C. J. Fritsch fil., G. D. Heuman, Ioan Stoercklin.

Editio denuo emendata et renovata. Folio (41 cm × 25½ cm). 1813.

Plates identical with those of earlier edition. Nos. 28–29, 33–34, 35–37, 39–40, 43–44, are on the same sheets folded to the size of the volume, but each plate has its individual number.

HALLIER, ERNST

DEUTSCHLANDS FLORA, ODER ABBILDUNG UND BESCHREIBUNG DER WILDWACHSENDEN PFLANZEN IN DER MITTEL-EUROPÄISCHEN FLORA

2 vols. 4to (28½ cm × 20½ cm). Dresden [1873–75].

Bound vols are also found with Leipzig on title-page.

500 unsigned hand-coloured engraved plates.

HANBURY, FREDERICK JANSON

AN ILLUSTRATED MONOGRAPH OF THE BRITISH HIERACIA.

8 parts Folio (38 cm × 28 cm) London 1889–98.

24 hand-coloured plates by Miss G. Lister and Mrs F. J. Hanbury, engraved by E. Carter; pp. 1–69 (pp. 33–36 never published), pls. 1–24 all published.

HARIOT, PAUL

LE LIVRE D'OR DES ROSES

4to (32 cm × 25 cm). Paris 1885.

60 chromolithographs. H.B.

HARTIG, THEODOR

VOLLSTÄNDIGE NATURGESCHICHTE DER FORSTLICHEN CULTURPFLANZEN DEUTSCHLANDS.

4to (27 cm × 22 cm). Berlin 1851.

120 hand-coloured engraved plates (Nos. 1–104, 35b–h, 37b–c, 41b–f, 45b–c) by F. Guimpel, engraved by C. Steglich, Guinand.

HARTINGER, ANTON

ENDLICHER'S PARADISUS VINDOBONENSIS, ABBILDUNGEN SELTENER UND SCHÖNBLÜHENDER PFLANZEN DER WIENER UND ANDERER GÄRTEN UND MUSEEN, VON ANTON HARTINGER . . . ERLÄUTERT VON BERTHOLD SEEMANN.

2 vols Folio (58 cm × 43 cm). Wien (Vienna) 1844–60.

81 chromolithographed plates (numbered 1–84, but 13–14, 41–42, 81–82 each represent one folded plates only) by A. Hartinger.

HAYNE, FRIEDRICH GOTTLOB & WILLDENOW, KARL LUDWIG

TERMINI BOTANICI ICONIBUS ILLUSTRATI ODER BOTANISCHE KUNSTSPRACHE DURCH ABBILDUNGEN ERLÄUTERT

Pinus Pinaster.

G. D. Ehret del. — Mackenzie sculp.

2 vols 4to (28 cm × 22 cm). Berlin 1807 (but actually published 1799–1812).
Coloured frontispiece, coloured title page, and 69 hand-coloured engravings. H.B.

HAYNE, FRIEDRICH GOTTLOB, & OTHERS

GETREUE DARSTELLUNG UND BESCHREIBUNG DER IN DER ARZNEIKUNDE GEBRÄUCHLICHEN GEWÄCHSE . . .
14 vols 4to (28 cm × 23 cm). Berlin 1805–37 [–48].
648 hand-coloured engraved and lithographed plates by F. Guimpel, W. Pape, J. Roeper and C. F. Schmidt, engraved by Franz, F. Guimpel, C. Haas, P. Haas, L. J. Haas and T. Weidlich, lithographed by C. F. Schmidt.
Vols. 1–13 have each 48 engraved plates, mostly by Guimpel. After Hayne's death in 1832, vol. 12 was completed by J. F. Brandt and J.T.C. Ratzeburg who then wrote vol. 13 and compiled a general index.
Vol. 14, never completed, with 24 lithographed hand-coloured plates by C. F. Schmidt and text by F. Klotzsch, was published in two parts; pls. 1–12 in 1843, pls. 13–24 in 1846.

HAYNE, FRIEDRICH GOTTLOB

ABBILDUNG DER FREMDEN, IN DEUTSCHLAND AUSDAUERNDEN HOLZARTEN.
4to (25 cm × 30 cm). Berlin 1825.
144 hand-coloured engraved plates by Friedrich Guimpel.

HEGETSCHWEILER, JOHANN, & LABRAM, JONAS DAVID

SAMMLUNG VON SCHWEIZER-PFLANZEN NACH DER NATUR UND AUF STEIN GEZEICHNET.
147 parts 8vo. Basel & Zurich [1827–34, 1838–46].
882 hand-coloured lithographed plates by J. D. Labram and L. Labram.
The first series comprised parts 1–80, each with 6 plates, making 480 plates, published 1827–34; the second series comprised parts 81–147, each with 6 plates, making 402 plates, published 1838–46; unfortunately the volumes are not dated and the plates not numbered. See Fr. Burckhardt in *Verhandl. Naturforsch. Ges. Basel* 19:1–36 (1907) for a biography and portrait of Labram and bibliographical notes on his work.

HEMPEL, GUSTAV & WILHELM, KARL

DIE BÄUME UND STRÄUCHER DES WALDES . . .
3 vols 4to (35 cm × 28 cm). Wien (Vienna) & Olmütz [1889–99].
60 chromolithographed plates, drawn by W. Liepolt, lithographed by A. Lenz and E. Beck; 342 text-figures.
According to E. M. Tucker in *J. Arnold Arb.* 2:237 (1921), Vol. 1, pp. 1–56 published in 1889, pp. 57–96 in 1890, pp. 97–128 in 1891, pp. 129–176 in 1892, pp. 153–200 in 1893; Vol. 2, pp. 1–16 in 1893, pp. 17–40 in 1894, pp. 41–88 in 1895, pp. 89–112 in 1896, pp. 113–148 in 1897; Vol. 3, pp. 1–48 in 1898, pp. 49–140 in 1899.

HEMSLEY, WILLIAM BOTTING

BIOLOGIA CENTRALI-AMERICANA, OR CONTRIBUTIONS TO THE KNOWLEDGE OF THE FAUNA AND FLORA OF MEXICO AND CENTRAL AMERICA, EDITED BY F. D. GODMAN AND O. SALVIN. BOTANY.
4 vols (text) and Atlas. 4to (31 cm × 24½ cm). London 1879–88.
109 plates (a few hand-coloured) drawn and lithographed by W. H. Fitch. 1 map (plate 110).

HENDERSON, EDWARD, & HENDERSON, ANDREW

THE ILLUSTRATED BOUQUET, CONSISTING OF FIGURES WITH DESCRIPTIONS OF NEW FLOWERS.
3 vols 4to [London] 1857–64
85 hand-colored engravings by James Andrews and C. T. Rosenberg, after the artwork of Miss Sowerby, Mrs Withers, and James Andrews

HENDERSON, PETER CHARLES

THE SEASONS OR FLOWER GARDEN, BEING A SELECTION OF THE MOST BEAUTIFUL FLOWERS THAT BLOSSOM AT THE FOUR SEASONS OF THE YEAR . . .
4to (36 cm × 29 cm). London 1806.
Frontispiece, coloured title-page and 24 hand-coloured engravings.
Henderson was one of the artists in Thornton's *Temple of Flora*. H.B.

HERBERT, WILLIAM

AN APPENDIX TO THE BOTANICAL REGISTER AND BOTANICAL MAGAZINE . . . CONTAINING A BOTANICAL ARRANGEMENT AND DESCRIPTION OF THE PLANTS HERETOFORE INCLUDED UNDER THE GENERA AMARYLLIS, CYRTANTHUS, CRINUM AND PANCRATIUM.
8vo (23½ cm × 15 cm). London 1821.

PLATE 35

"Pinus Pinaster" illustrates Aylmer Lambert's monograph,
A Description of the Genus Pinus *(London 1803–24). The plates are acclaimed for their three-dimensional depiction of pine needles. Daniel Mackenzie engraved this print after a watercolor by Georg Ehret [q.v.] owned by Sir Joseph Banks, a great patron of botany. The realism of Lambert's book moved Goethe to exclaim: "It is a real joy to look at these plates, for nature is revealed, art concealed."*

2 hand-coloured plates, one lithographed and engraved, and two uncoloured folded lithographed plates, all by Herbert, lithographed by Herbert or engraved by J. Watts.
An important contribution to the taxonomy of the Amaryllidaceae, issued with the index to vols. 43–48 (1815–21) of the *Botanical Magazine* and usually bound in vol. 48 of the *Botanical Register*. The title given above is taken from the original wrapper. Probably published in December 1821: see W. T. Stearn in *J. Soc. Bibl. Nat. Hist.* 2:376 (1952).

HERBERT, WILLIAM
AMARYLLIDACEAE; PRECEDED BY AN ATTEMPT TO ARRANGE THE MONOCOTYLEDONOUS ORDERS, AND FOLLOWED BY A TREATISE ON CROSS-BRED VEGETABLES, AND SUPPLEMENT.
8vo (25 cm × 15½ cm). London 1837.
48 plates by Herbert, engraved by H. H. Weddell.
This exists with all the plates uncoloured, or most of them hand-coloured.
As regards the colouring Herbert stated 'the publisher having been desirous of offering some copies to the public with coloured plates, it must be understood that, where no live specimen has been seen in this country, and no precise memorandum of the colours has been given by the collector, the plates could only be made to represent the existing tints of the dried specimens, which in many cases are very fallacious'.
This accounts for the misleading dull yellowish colour of some figures.
Probably published at the end of April 1837; see W. T. Stearn in *J. Soc. Bibl. Nat.* Hist. 2: 376 (1952).

HERINCQ, F. See periodicals, L'HORTICULTEUR FRANÇAIS.

HIBBERD, JAMES SHIRLEY
NEW AND RARE BEAUTIFUL-LEAVED PLANTS, CONTAINING ILLUSTRATIONS AND DESCRIPTIONS OF THE MOST ORNAMENTAL-FOLIAGED PLANTS NOT HITHERTO NOTICED IN ANY WORK ON THE SUBJECT.
8vo (25 cm × 17 cm). London 1868.
54 coloured wood-engravings by A. F. Lydon.
Reissued in 1870.

HIBBERD, JAMES SHIRLEY
THE IVY, A MONOGRAPH . . .
4to (21 cm × 16½ cm). London 1872.
66 wood-engraved figures in the text and 4 coloured wood-engravings by A. Slocombe.
Of more horticultural than artistic merit but an interesting example of ornate Victorian book production with ivy-leaf borders to the text.

HILL, JOHN
THE BRITISH HERBAL: AN HISTORY OF PLANTS AND TREES NATIVES OF BRITAIN.
Folio (41 cm × 22 cm). London 1756.
Coloured frontispiece, and 75 hand-coloured engravings. H.B.

HILL, JOHN
THE VEGETABLE SYSTEM . . . WITH FIGURES . . . FROM NATURE ONLY.
26 vols Folio (46 cm × 28½ cm). London [1759]–1775.
1546 hand-coloured engraved plates, all by John Hill.
In most sets vols 1 and 2 are represented by a second edition of 1770–73; between 1772 and 1775, vols 3–21 were also re-issued. The plates are as follows: Vol. 1, pls 1–21; vol. 2, pls 1–87, 26 bis; vol. 3, pls 87 bis–137; vol. 4, pls 1–46; vol. 5, pls 1–53; vol. 6, pls 1–62; vol. 7, pls 1–60; vol. 8, pls 1–60; vol. 9, pls 1–60; vol. 10, pls 1–59, 44 bis; vol. 11, pls 1–60; vol. 12, pls 1–60 + 35 bis, 1–10 (in 1773 ed. 1–70 + 35 bis); vol. 13, pls 1–61, + 11-2- (in 1773 ed. 1–61 +71–80); vol. 14, pls 1–60; vol. 15, pls 1–61; vol. 16, pls 1–61; vol. 17, pls 1–60; vol. 18, pls 1–60; vol. 19, pls 1–60; vol. 20, pls 1–60; vol. 21, pls 1–60; vol. 22, pls 1–61; vol. 23, pls 1–60; vol. 24, pls 1–60; vol. 25, pls 1–60; vol. 26, pls 1–60.

HILL, JOHN
EDEN OR A COMPLEAT BODY OF GARDENING.
Folio (43 cm × 24 cm). London 1757.
Coloured frontispiece and 60 hand-coloured engravings.
2nd ed., 1773 with 80 plates. H.B.

HILL, JOHN
EXOTIC BOTANY, ILLUSTRATED IN 35 FIGURES OF CURIOUS AND ELEGANT PLANTS . . .
Folio (46½ cm × 29 cm). London 1759.
35 engraved plates by J. Hill.
Ed. 2. London. 1772. This has 'of Chinese and American Shrubs and Plants' in the title but is otherwise the same.

HILL, JOHN
A DECADE OF CURIOUS AND ELEGANT TREES AND PLANTS, DRAWN AFTER SPECIMENS RECEIVED FROM THE EAST INDIES AND AMERICA.
Folio (53 cm × 36 cm). London 1773.
10 hand-coloured engraved plates by John Hill.
DECADE DI ALBERI CURIOSI ED ELEGANTI PIANTE DELLE INDIE ORIENTALI E DELL' AMERICA.
8vo. (24 cm × 18 cm). Roma (Rome) 1786.
10 unnumbered hand-coloured engraved plates copied by Majoli, engraved by Bianchi and Majoli.

HILL, JOHN

TWENTY-FIVE NEW PLANTS, RAIS'D IN THE ROYAL GARDENS AT KEW; THEIR HISTORY AND FIGURES.
Folio (29½ cm × 47½ cm). London 1773.
25 unsigned engraved plates.

HOFFMANN, GEORG FRANZ

HISTORIA SALICUM ICONIBUS ILLUSTRATA.
2 vols Folio (37 cm × 23½ cm). Lipsiae (Leipzig) 1785–91.
30 plates (pls. 1–29, 31) by G. F. Hoffmann, engraved by J. Nussbiegel and Capieux. Vol. 1, fasc. 1–2, pp. 1–48, pls 1–10 in 1785, fasc. 3, pp. 49–66, pls 11–16 in 1786, fasc. 4, pp. 67–78, pls 17–24 in 1787; vol. 2, fasc. 1, pp. 1–12, pls 25–31 (30 never published) in 1791.

HOFFMANNSEGG, JOHANN CENTURIUS GRAF VON, & LINK, HEINRICH FRIEDRICH

FLORE PORTUGAISE OU DESCRIPTION DE TOUTES LES PLANTES QUI CROISSENT NATURELLEMENT EN PORTUGAL . . .
2 vols. Folio (53 cm × 34½ cm). Berlin 1809–20 [–40].
114 engraved and lithographed plates (Nos 1–109, 90b, 108b, and 3 unnumbered planches d'instruction) of which all but the unnumbered plates are partly colour-printed, partly hand-coloured, by Hoffmannsegg and G. W. Völcker (Voelker), engraved by F. W. Bollinger, A. Clar, C. Dümbte, F. Guimpel, M. and P. Haas, J. F. Krethlow, J. D. Laurens, F. W. Meyer, L. Schubert and Wachsmann.
This work was published in parts and never completed. The contents and dates of publication of the parts are imperfectly known. It seems probable that publication was as follows: vol. 1, parts 1–10, pls 1–50 in 1809–13, parts 11–16, pls 51–78 in 1813–20; vol. 2, parts 17–21, pls 79–102 in 1820–28, part 22, pls 103–106 in 1833, part 23, pls 107–109 in 1840. The text ends with p. 436.

HOGG, ROBERT, & BULL, HENRY GRAVES

THE HEREFORDSHIRE POMONA, CONTAINING ORIGINAL FIGURES AND DESCRIPTIONS OF THE MOST ESTEEMED KINDS OF APPLES AND PEARS.
2 vols 4to (39 cm × 29½ cm). Hereford & London 1878–85.
76 chromolithographed plates by Edith E. Bull, Alice B. Ellis and W. G. Smith (pl. 1 only), lithographed by G. Severeyns. Published under the auspices of the Woolhope Club. Although the title-page is dated '1876–1885' the wrapper of part 1 is dated '1878'.
One of the finest fruit books ever issued.

HOOKER, JOSEPH DALTON

THE BOTANY OF THE ANTARCTIC VOYAGE OF H.M. DISCOVERY SHIPS EREBUS AND TERROR IN THE YEARS 1839–43 . . .

1. FLORA ANTARCTICA
2 vols 4to (30½ cm × 24 cm). London [1844–] 1847.
187 hand-coloured plates (pls. 1–198, of which 9–10, 22–23, 24–25, 26–27, 31–32, 39–40, 44–45, 90–91, 136–137, 165–166, represent one plate each, 167 bis and 167 ter) by W. H. Fitch, J. D. Hooker and W. W., lithographed by W. H. Fitch. For dates of publication, see F. G. Wiltshear in *J. Bot. (London)* 51: 357 (1913), *Flora Malesiana* I. 4: clxxxvii (1854).

2. FLORA NOVAE-ZELANDIAE.
2 vols. 4to. (30½ cm × 24 cm). London. [1852–] 1853–55.
127 hand-coloured plates (pls 1–130, of which 54–55, 59–60 and 114–115 represent one plate each) by M.J.B., W. Fitch, W.H.H., W. Mitten, and W. Wilson, lithographed by W. Fitch and W.H.H.

3. FLORA TASMANIAE.
2 vols. 4to (30 cm × 24 cm). London. 1855–59.
199 hand-coloured plates (Nos 1–200 of which 84–85 represents one plate only) by W. Archer, M. J. Berkeley, W. H. Fitch, W. H. H., W. Mitten, Wilson, lithographed by W. H. Fitch and W.H.H. For dates of publication, see above.

HOOKER, JOSEPH DALTON

THE RHODODENDRONS OF SIKKIM-HIMALAYA . . . FROM DRAWINGS AND DESCRIPTIONS MADE ON THE SPOT . . .
3 parts Folio (50 cm × 36 cm). London 1849 [–51].
30 hand-coloured lithographed plates by J. D. Hooker, lithographed by W. H. Fitch.
Part 1, pls 1–10 issued in 1849; parts 2 and 3, pls 11–30 in 1851.
An important work both for the botanist and horticulturist since it contains descriptions and plates of many of the best garden Rhododendron species which can be grown in this country and an account of their discovery. Fitch's lithographs are made from J. D. Hooker's field sketches. P.M.S.

HOOKER, JOSEPH DALTON

ILLUSTRATIONS OF HIMALAYAN PLANTS, CHIEFLY SELECTED FROM DRAWINGS MADE FOR J. F. CATHCART . . . THE DESCRIPTIONS AND ANALYSES BY J. D. HOOKER . . . THE PLATES EXECUTED BY W. H. FITCH.
Folio (50 cm × 36½ cm). London 1855.
24 hand-coloured lithographed plates by W. H. Fitch and J. D. Hooker, based on drawings by Indian artists, lithographed by W. H. Fitch.
Published in July 1855 according to Stearn in *Calcutta R. Bot. Garden*, 150th Anniv. Vol. 115 (1942). *Contains probably the finest plates of Magnolia Campbellii and Meconopsis simplicifolia ever made as well as other important Himalayan plants.* P.M.S.

Passiflora quadrangularis. Grenadille quadrangulaire.

Poiteau Pinxit | P. Duménil Direxit | Robert Sculpsit

HOOKER, WILLIAM

POMONA LONDINENSIS: CONTAINING COLOURED ENGRAVINGS OF THE MOST ESTEEMED FRUITS CULTIVATED IN THE BRITISH GARDENS

8 parts 4to (40½ cm × 32 cm). London [1813–] 1818.
49 colored aquatint engraved plates by W. Hooker.

HOOKER, WILLIAM. See SALISBURY, R. A. 1805.

HOOKER, SIR WILLIAM JACKSON

EXOTIC FLORA, CONTAINING FIGURES AND DESCRIPTIONS OF NEW, RARE OR OTHERWISE INTERESTING EXOTIC PLANTS . . .

3 vols 8vo (24 cm × 15 cm). Edinburgh [1822–] 1823–27.
233 hand-coloured plates (pls 1–232 & 163*) by R. K. Greville, L. Guilding, J. Lindley and A. Menzies engraved by J. Swan. For dates of publication, see W. T. Stearn in *Flora Malesiana* I. 4: clxxvii (1954).

HOOKER, SIR WILLIAM JACKSON

BOTANICAL MISCELLANY; CONTAINING FIGURES AND DESCRIPTIONS OF SUCH PLANTS AS RECOMMEND THEMSELVES BY THEIR NOVELTY, RARITY OR HISTORY, OR BY THE USES TO WHICH THEY ARE APPLIED . . .

3 vols 8vo (24½ cm × 15 cm). London [1829–] 1830–33.
153 engraved plates (Nos. 1–112, 67 bis, and Supplementary plates 1–19, 21–41) by Dr Greville, W. J. Hooker, L. Guilding, A. M., W. Wilson, J. Newman, Mme. Bernard, Prof. Boyer, J. F. Klotzsch, Mrs C. Telfair, Miss Baugrie, engraved by Swan.

HOOKER, SIR WILLIAM JACKSON

FLORA BOREALI-AMERICANA; OR THE BOTANY OF THE NORTHERN PARTS OF BRITISH AMERICA . . .

2 vols 4to (29 cm × 23½ cm). London [1829–] 1833–40.
238 plates by R. K. Greville and W. J. Hooker, engraved by J. Swan.
1 map. For dates of publication, see M. J. van Steenis-Kruseman & W. T. Stearn in *Flora Malesiana* I. 4: clxxxviii (1954).

HOOKER, SIR WILLIAM JACKSON

THE BOTANY OF CAPT. BEECHEY'S VOYAGE, COMPRISING AN ACCOUNT OF THE PLANTS COLLECTED BY . . . OFFICERS OF THE EXPEDITION DURING THE VOYAGE TO THE PACIFIC AND BEHRING'S STRAIT PERFORMED IN H.M.S. BLOSSOM . . . IN THE YEARS 1825–28.

4to (26½ cm × 21 cm). London [1830–] 1841.
100 uncoloured plates (Nos 1–99, 29 bis) by W. J. Hooker, W. P. Beechey and W. H. Fitch, engraved by Swan.
For dates of publication, see van Steenis-Kruseman & W. T. Stearn in *Flora Malesiana* I. 4: clxxxviii.

HOOKER, SIR WILLIAM JACKSON

A CENTURY OF ORCHIDACEOUS PLANTS SELECTED FROM CURTIS'S BOTANICAL MAGAZINE . . .

4to (31 cm × 24 cm). London 1846.
100 hand-coloured lithographed plates all by W. H. Fitch.
Another issue. 1851.

HOOKER, SIR WILLIAM JACKSON

DESCRIPTION OF VICTORIA REGIA, OR GREAT WATER-LILY OF SOUTH AMERICA

Folio (39½ cm × 25½ cm). London 1847.
4 hand-coloured lithographed plates by W. H. Fitch.

HOOKER, SIR WILLIAM JACKSON

VICTORIA REGIA; OR ILLUSTRATIONS OF THE ROYAL WATER LILY . . .

Folio (74 cm × 54 cm). London 1851.
4 hand-coloured lithographed plates by W. H. Fitch.
A larger and more detailed work than the above, with very impressive plates specially drawn and not copied from it.

HOOLA VAN NOOTEN, BERTHE

FLEURS, FRUITS ET FEUILLAGES CHOISIES DE LA FLORE ET DE LA POMONE DE L'ILE DE JAVA, PEINTS D'APRÉS NATURE.

Folio (58 cm × 44 cm). Bruxelles (Brussels) 1863–64.
40 chromolithographed plates by B. Hoola van Nooten, lithographed by G. Severeyns.
Ed. 2. 1880. 40 chromolithographed plates as above.
Ed. 3 1885. 40 chromolithographed plates, lithographed by P. Depannemacker.

HOST, NICOLAUS THOMAS

ICONES ET DESCRIPTIONES GRAMINUM AUSTRIACORUM

4 vols Folio (46½ cm × 32 cm). Vindobonae (Vienna) 1801–09.
400 hand-coloured engraved plates, unnumbered and unsigned but stated by Host in vol. 4:56 to be all (except vol. 4 pl. 38, *Arundo Donax*) by 'Johanes Babtista Jebmayer, vir artis suae peritissimus' (J. Ibmayer).
Vol. 1, 1801; vol. 2, 1802; vol. 3, 1805; vol. 4, 1809.

PLATE 36
Pierre Poiteau drew "Passiflore quadrangularis" while botanizing in the French West Indies. Robert etched the design in roulette and line to illustrate F. Richard de Tussac's Flore des Antilles *(Paris 1808–27). The plate was inked in blue which blends unobtrusively with the hand-coloring in green and red.*

HOST, NICOLAUS THOMAS

SALIX . . .

Vol. 1 (all published). Folio (45½ cm × 33 cm). Vindobonae (Vienna) 1828 [–30].

105 hand-coloured engraved plates unsigned but by J. Ibmayer.

HUCHES, GRIFFITH

THE NATURAL HISTORY OF BARBADOS.

Folio (41½ cm × 26½ cm). London 1750.

30 uncoloured plates (pls. 1–29, 10*) of which 22 are botanical, by Ehret and G. Bickham, engraved by Austen, G. Bickham and Mynde.

HUMBOLDT, FRIEDRICH HEINRICH ALEXANDER VON, & BONPLAND, AIMÉ

VOYAGE . . . AUX RÉGIONS ÉQUINOXIALES DU NOUVEAU CONTINENT. PARTIE 6, BOTANIQUE.

This important work consists of 6 sections which for convenience in citation are treated as independent works. For dates of publication, see C. D. Sherborn and B. B. Woodward in *J. Bot.* (London) 39: 202–205 (1901), J. H. Barnhart in *Bull. Torrey Bot. Club* 29: 585–598 (1902). For place-names cited, see T. A. Sprague, 'Humboldt and Bonpland's Itinerary in Mexico', *Kew Bull.* 1926: 23–30, and N. Y. Sandwith, 'Humboldt and Bonpland's Itinerary in Venezuela', *Kew Bull.* 1925: 295–310.

Sect. 1 Plantes équinoxiales, See Bonpland, A. 1805,

Sect. 2 Monographie des Mélastomacées. See Bonpland, A. (1806–) 1816.

Sect. 3 Nova Genera et Species. See Kunth, C. S. 1815.

Sect. 4 Mimoses. See Kunth, C. S. 1819

Sect. 5 Synopsis Plantarum. 4 vols. without plates. Paris 1822–26.

Sect. 6 Revision des Graminées. See Kunth, C. S. 1829.

ITO, KEISUKE

KOISHIKAWA SHOKUBUTSUYEN SŌMOKU DZUSETSU. FIGURES AND DESCRIPTIONS OF PLANTS IN KOISHIKAWA BOTANICAL GARDEN.

2 vols Folio (40½ cm × 28 cm). Tokio 1883–84

52 hand-coloured plates of which those in vol. 1 (pls 1–19, 2b and 2c) are engraved, and those in vol. 2 (pls 1–28, 8b, 11b, 25b) are lithographed.

Title-page and numbering of vol. 1 in Japanese only.

JACQUEMONT, VICTOR

VOYAGE DANS L'INDE . . . PENDANT LES ANNÉES 1828–32.

4 vols and Atlas, 2 vols 4to. (35 cm × 26½ cm). Paris [1835–] 1841–44.

Vol. 2 of Atlas contains 180 botanical engraved plates by E. Delile, A. Riocreux and Borromée.

JACQUIN, JOSEPH FRANZ VON

ECLOGAE PLANTARUM RARIORUM AUT MINUS COGNITARUM . . .

2 vols Folio (47 cm × 32½ cm). Vindobonae (Vienna) 1811–44.

167 hand-coloured plates (pls. 1–100 and a supplementary plate, 101–156 of which 153–4 and 155–6 represent one plate each, 158–169) engraved by Küka and J. Seher. No plate 167. Vol. 1, pls 1–100, published 1811–16 (pls 41–60 in 1813). Vol. 2, pls 101–169 edited by E. Fenzl, in 1844.

JACQUIN, JOSEPH FRANZ VON

ECLOGAE GRAMINUM RARIORUM AUT MINUS COGNITARUM . . .

5 parts Folio (46 cm × 33½ cm). Vindobonae (Vienna) 1813–40.

48 hand-coloured engraved plates.

Issued in 5 parts, the fifth being extremely rare: fasc. 1, pp. 1–16, pls 1–10, in 1813; fasc. 2, pp. 11–32, pls 11–20; fasc. 3–4, pp. 33–64, pls 21–40 in 1820; fasc. 5 (edited by Fenzl) p. 65, pls 41–45, 47–49 in 1840.

JACQUIN, NICOLAUS JOSEPH VON

OBSERVATIONUM BOTANICARUM ICONIBUS AB AUCTORE DELINEATIS ILLUSTRATARUM.

Folio (36 cm × 24 cm). Vindobonae (Vienna) 1764–71.

100 coloured plates by N. J. Jacquin, engraved by Wangner.

JACQUIN, NICOLAUS JOSEPH VON

HORTUS BOTANICUS VINDOBONENSIS SEU PLANTARUM RARIORUM, QUAE IN HORTO BOTANICO VINDOBONENSIS . . . COLUNTUR . . .

3 vols Folio (48 cm × 28 cm). Vindobonae (Vienna) 1770–76.

300 hand-coloured, unsigned engraved plates by F. Scheidl.

JACQUIN, NICOLAUS JOSEPH VON

FLORAE AUSTRIACAE, SIVE PLANTARUM SELECTARUM IN AUSTRIAE ARCHIDUCATU SPONTE CRESCENTIUM, ICONES . . .

5 vols Folio (46 cm × 28 cm). Vienna 1773–78.

500 hand-coloured unsigned engraved plates (pls 1–450, app. 1–50) mostly by F. Scheidl, engraved by I. Adam. *Very fine plates.* W.B.

JACQUIN, NICOLAUS JOSEPH VON

MISCELLANEA AUSTRIACA AD BOTANICAM, CHEMIAM, ET HISTORIAN NATURALEM SPECTANTIA . . .

2 vols 4to (22cm × 17 cm). Vindobonae (Vienna) 1778–81.

44 hand-coloured unsigned plates (Vol. 1, pls. 1–21;

Vol. 2, pls. 1–23) by Jacquin, engraved by J. Adam. According to B. G. Schubert in *Contr. Gray Herb.* 154: 3–5 (1945), vol. 1, pls. 1–21, probably not issued before 1779, vol. 2, pls. 1–23 probably late 1781 or early 1782.

JACQUIN, NICOLAUS JOSEPH VON

SELECTARUM STIRPIUM AMERICANARUM HISTORIA . . . ADJECTIS ICONIBUS AD AUTORIS ARCHETYPA PICTIS.

Folio (46 cm × 32 cm). [Vindobonae (Vienna) c. 1780.]

264 hand-drawn and coloured plates by N. J. Jacquin. At most only 18 copies (of which uncoloured engraved reproductions were published in Jacquin's *Select. Stirp. Amer.* of 1783), and much reduced hand-coloured engraved 8vo reproductions in Zorn's *Dreyhundert auserlesene amerikanische Gewächse* of 1785–89. The plates are directly copied from Jacquin's originals without any engraved outlines. Copies exist in London (Kew; Lindley Library; British Museum in Bankseian and King's Libraries; British Museum (Nat. Hist.), Haarlem, Cambridge, Mass., Washington, D.C. and New York; others are recorded by F. Wilshear in *J. Bot (London), 51; 140* (1913) as being then in Vienna, Berlin and Dresden; another one is known to be in private hands).

Plates of great botanical importance though artistically worthless. W.B.

JACQUIN, NICOLAUS JOSEPH VON

ICONES PLANTARUM RARIORUM . . .

3 vols Folio (47 cm × 29 cm). Vindobonae (Vienna) 1781–93.

648 hand-coloured unsigned unnumbered engraved plates, mainly by J. Hofbauer, J. Scharf, Ferd. Bauer and Fr. Bauer, engraved by J. Adam. For dates of publication, see B. G. Schubert in *Contr. Gray Herb.* 154: 3–23 (1945). For details of artists, see preface to Jacquin, *Pl. rar. Horti Caes. Schoenbr.* reprinted in Nissen, *Bot. Buchill.* 2:91 (1951).

JACQUIN, NICOLAUS JOSEPH VON

COLLECTANEA AUSTRIACA AD BOTANICAM, CHEMIAM ET HISTORIAM NATURALEM SPECTANTIA.

5 vols (including Supplement) 4to (29 cm × 23 cm). Vindobonae (Vienna) 1786–96.

106 hand-coloured unsigned engraved plates by Jacquin. (Vol. 1, pls. 1–22 in 1786 or 87; vol. 2, pls. 1–18 in 1789; vol. 3, pls. 1–23 in 1790; vol. 4, pls. 1–27 in 1791; vol. 5, pls. 1–16 in 1796). For dates of publication see B. G. Schubert in *Contr. Gray Herb.* 154 3–5 (1945).

JACQUIN, NICOLAUS JOSEPH VON

OXALIS, MONOGRAPHIA ICONIBUS ILLUSTRATA.

4to (27½ cm × 23½ cm). Viennae 1794.

81 engraved plates (of which 75 are hand-coloured) by J. Scharf.

JACQUIN, NICOLAUS JOSEPH VON

PLANTARUM RARIORUM HORTI CAESAREI SCHOENBRUNNENSIS DESCRIPTIONES ET ICONES.

4 vols Folio (48 cm × 34 cm). Viennae 1797–1804.

500 hand-coloured unsigned engraved plates by J. Scharf and M. Sedelmayer.

JACQUIN, NICOLAUS JOSEPH VON

STAPELIARUM IN HORTIS VINDOBONENSIBUS CULTARUM DESCRIPTIONES FIGURIS COLORATIS ILLUSTRATAE.

Folio (45 cm × 33 cm). Vindobonae (Vienna) 1806 [–1819].

64 unnumbered hand-coloured engraved plates.

JACQUIN, NICOLAUS JOSEPH VON

FRAGMENTA BOTANICA, FIGURIS COLORATIS ILLUSTRATA . . .

Folio (52 cm × 36 cm). Viennae 1809.

138 hand-coloured unsigned engraved plates.

JAUBERT, HIPPOLYTE FRANÇOIS, COMTE DE, & SPACH, EDOUARD

ILLUSTRATIONES PLANTARUM ORIENTALIUM (OU CHOIX DE PLANTES NOUVELLES OU PEU CONNUES DE L'ASIE OCCIDENTALE).

5 vols 4to (35 cm × 26 cm). Paris 1842–57.

500 uncoloured plates by C. Aubriet, Mlle Champeau, J. Decaisne (analyses), J. Gontier, Mme Hublier, E. Lesèble, de Ligneville, Maubert, P. J. Redouté, Riocreux, N. Robert, Mme Spach, Willy, engraved by many different engravers.

The parts of the title given above in brackets appeared only on the wrappers. Plates by Aubriet, Redouté and Robert are based on coloured drawings of the collection of Vélins of the Muséum national d'Histoire naturelle, Paris. For dates of publication see W. T. Stearn in *J. Soc. Bibl. Nat. Hist.* 1: 255–259 (1939), summarised in *Flora Malesiana* I, 4(5), cxci (1954).

JAUME SAINT-HILAIRE, JEAN HENRI

PLANTES DE LA FRANCE DÉCRITES ET PEINTES D'APRES NATURE.

10 vols 4to. (26 cm × 17½ cm). Paris [1805–] 1808–22.

1000 coloured and engraved plates by Jaume Saint-Hilaire, engraved by Dubreuil and Veron.

Another issue, Paris 1819–22.

Charming little stipple plates. W.B.

JAUME SAINT-HILAIRE, JEAN HENRI

LA FLORE ET LA POMONE FRANÇAISES, OU HISTOIRE ET FIGURES EN COULEURS DES FLEURS ET DES FRUITS DE FRANCE.

6 vols Folio (42 cm × 25 cm). Paris 1828–33.

544 colour-printed unsigned engraved plates by Jaume Saint-Hilaire.

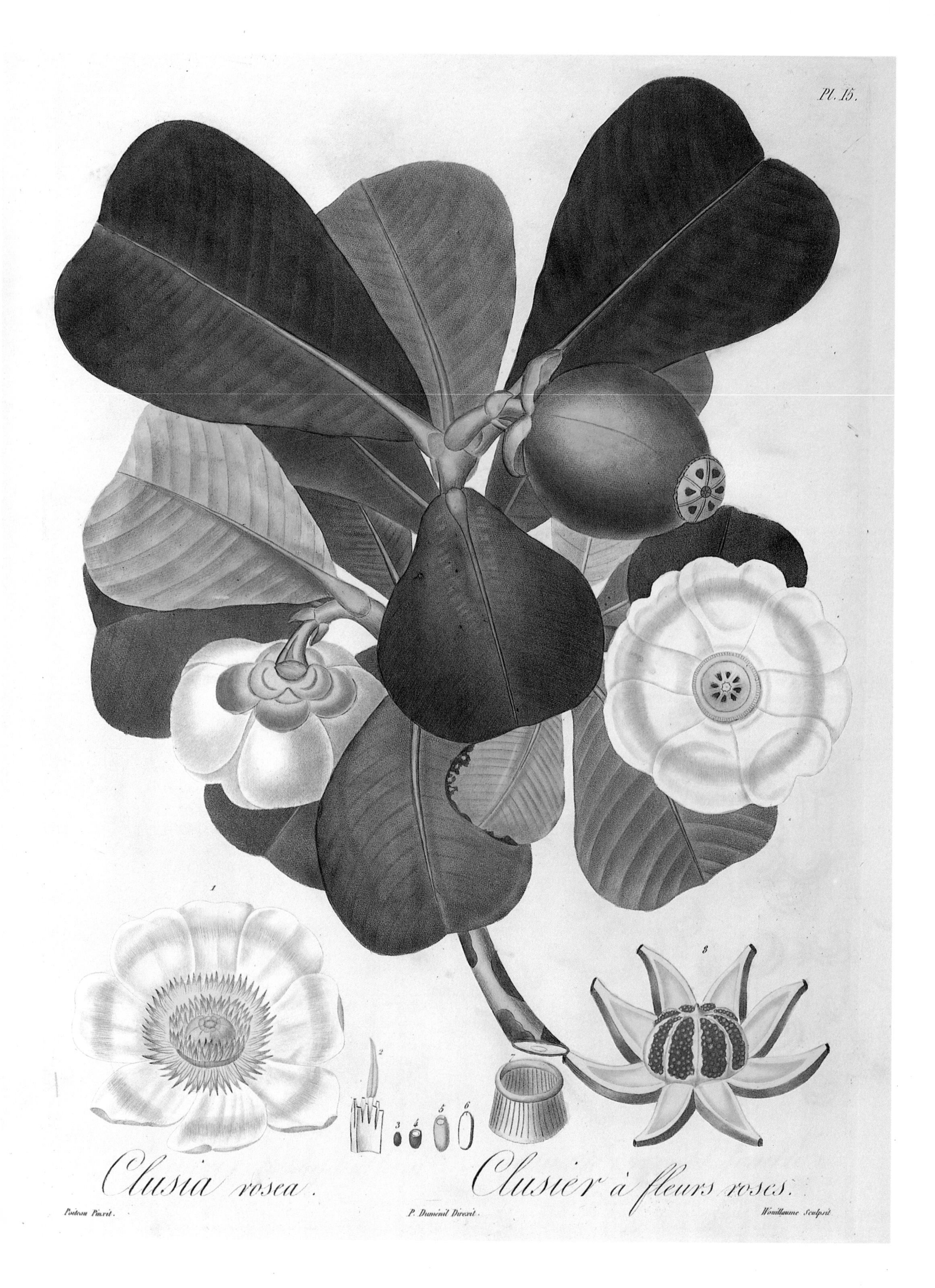
Pl. 15.
1
2
3
4
5
6
7
8
Clusia rosea.
Clusier à fleurs roses.
Poiteau Pinxit.
P. Duménil Direxit.
Wouillaume Sculpsit

JOLYCLERC, C. NICOLAS

PHYTOLOGIE UNIVERSELLE OU HISTOIRE NATURELLE ET METHODIQUE DES PLANTES.

8 vols 8vo (20 cm × 12 cm). Paris An. VII (1799).

638 hand-coloured engravings. Pritzel 4466 (inaccurate). H.B.

JORDAN, ALEXIS & FOURREAU, JULES

ICONES AD FLORAM EUROPAE NOVO FUNDAMENTO INSTAURANDAM SPECTANTES.

3 vols Folio (37 × 30 cm). Parisiis (Paris) 1866–1903.

501 hand-coloured engraved plates (pls. 1–500, 335 bis, 337 bis, 460 missing) by C. Delorme, J. Fourreau, P. Galland and A. Mignot, engraved by Charbonnier and P. Galland.

For dates of publication see W. T. Stearn in *Flora Malesiana* I. 4: cxcii (1954). Publication stopped at Vol. 2, p. 24, pl. 280 in 1871 and was not resumed until 1903.

KAEMPFER, ENGELBERT

AMOENITATUM EXOTICARUM POLITICO-PHYSICO-MEDICARUM FASCICULI, QUIBUS CONTINENTUR VARIAE RELATIONES, OBSERVATIONES ET DESCRIPTIONES RERUM PERSICARUM ET ULTERIORIS ASIAE MULTA ATTENTIONE, IN PEREGRINATIONIBUS PER UNIVERSUM ORIENTEM, COLLECTAE.

4to (22 cm × 18 cm). Lemgoviae (Lemgo) 1712.

Frontispiece and 90 uncoloured plates by Kaempfer, engraved by F. W. Brandshagen, Donop and J. Gole.

Kaempfer was a vigorous and accurate botanical draughtsman. W.B.

An important work for students of Japanese plants, giving the first descriptions of many well-known garden plants such as *Ginkgo biloba* and *Camellia japonica.*

A travel book, but with pp. 765–912 devoted to the flora of Japan; see W. T. Stearn in R. Hort. Soc., *Lily Year Book* 12: 65–70 (1948).

KAEMPFER, ENGELBERT

ICONES SELECTAE PLANTARUM QUAS IN JAPONIA COLLEGIT ET DELINEAVIT E. KAEMPFER; EX ARCHETYPIS IN MUSEO BRITANNICO ASSERVATIS EDIDIT. JOS. BANKS.

Folio (43 cm × 26½ cm). Londini (London) 1791.

59 plates engraved by Mackenzie from Kaempfer's original drawings in the British Museum; cf. B. Henry in *J. Soc. Bibl. Nat. Hist.* 3: 104 (1955). For modern botanical names, see T. Nakai in *J. Arnold Arb.* 6: 186–188 (1925). A few original drawings of Lilium are reproduced by W. T. Stearn in *R. Hort. Soc. Lily Year Book* 12: figs 26–29 (1948) and of Hosta by N. Hylander in *Acta Horti Berg.* 16: 350–351 (1954).

KARSTEN, KARL WILHELM GUSTAV HERMANN

AUSWAHL NEUER UND SCHÖNBLÜHENDER GEWÄCHSE VENEZUELA'S.

2 Parts 4to (29 cm × 22½ cm). Berlin 1848.

12 hand-coloured lithographed plates all by C. F. Schmidt.

KARSTEN, KARL WILHELM GUSTAV HERMANN

FLORAE COLUMBIAE TERRARUMQUE ADJACENTIUM SPECIMINA SELECTA IN PEREGRINATIONE DUODECIM ANNORUM OBSERVATA.

2 vols. Folio (50 cm × 34 cm). Berolini (Berlin) 1858–69.

200 hand-coloured plates by Düwel, C. F. Schmidt, F. Wagen, H. Karsten, V. Erbe, Chodowiecki, Klein, Borchel, F. Hecht, H. Troschel, C. F. Appun, lithographed by Düwel, F. Wagner, Chodowiecki and W. A. Meyn.

[KER, CHARLES HENRY BELLENDEN].

ICONES PLANTARUM SPONTE CHINA NASCENTIUM, E BIBLIOTHECA BRAAMIANA EXCERPTA.

Folio (49 cm × 34 cm). London 1821.

30 hand-coloured plates lithographed (23) or etched on copper (7) by C.H.B. Ker, being copies of Chinese drawings brought from China by A. E. van Braam-Houckgeest. They are not numbered and have no names; identifications arranged according to the sequence of the Kew copy are provided by W. B. Hemsley in E. Bretschneider, *Hist. Europ. Discov. in China* 1:187 (1898). J. Lindley probably wrote the preface. It appears that this work was first issued in 1818 with the title *Icones pictae Indo-Asiaticae Plantarum excerptae ex Codicibus Dom. Cattley.*

KERNER, JOHANN SIMON

BESCHREIBUNGEN UND ABBILDUNGEN DER BÄUME UND GESTRÄUCHE, WELCHE IN DEM HERZOGTHUM WIRTEMBERG WILDWACHSEN.

1 vol. Parts 1–9. 4to (25½ cm × 20½ cm). Stuttgart 1783–92.

71 hand-coloured engraved plates by J. S. Kerner.

KERNER, JOHANN SIMON

ABBILDUNGEN ALLER OEKONOMISCHEN PFLANZEN. FIGURES DES PLANTES ÉCONOMIQUES.

PLATE 37

"Clusia rosea," a West Indian plant, was drawn from life by Pierre Poiteau for F. Richard de Tussac's Flore des Antilles *(Paris 1808–27). Wouillaume etched this print in stipple, and it was inked in four colors for a botanically accurate and beautiful illustration. It is named for Clusius, the sixteenth-century Swiss botanist who pioneered the study of New World plants.*

8 vols 4to (27½ cm × 22 cm). Stuttgart 1786–96.
800 hand-coloured etched plates by J. S. Kerner.
Delicately etched plates. W.B.

KERNER, JOHANN SIMON
DARSTELLUNG VORZÜGLICHER AUSLÄNDISCHER BÄUME UND GESTRÄUCHE, WELCHE IN DEUTSCHLAND IM FREIEN AUSDAUERN.
1 vol Parts 1–4. 4to (28 cm × 22 cm). Tübingen 1796.
60 hand-coloured engraved plates by J. S. Kerner.

KERNER, JOHANN SIMON
HORTUS SEMPERVIRENS EXHIBENS ICONES PLANTARUM SELECTIORUM . . .
71 parts Folio (64 cm × 47½ cm). Stuttgart 1796–1830.
851 original water-colour drawings by J. S. Kerner, without engraved outlines, most of them copied or compounded from those in other publications, e.g. *Botanical Magazine*, Jacquin's works, etc. Complete sets of this work are extremely rare.
A work of gallant but misdirected energy. W.B.

KLOTZSCH, JOHANN FRIEDRICH, & GARCKE, AUGUST
DIE BOTANISCHEN ERGEBNISSE DER REISE DES PRINZEN WALDEMAR VON PREUSSEN IN DEN JAHREN 1845 UND 1846. DURCH WERNER HOFFMEISTER AUF CEYLON, DEM HIMALAYA UND AN DEN GRENZEN VON TIBET GESAMMELTE PFLANZEN BESCHREIBEN VON F. KLOTZSCH UND A. GARCKE.
4to (35 cm × 26½ cm). Berlin 1862.
100 plates by C. Müller and C. F. Schmidt, lithographed by C. F. Schmidt.

KLOTZSCH, JOHANN FRIEDRICH. See LINK, H. F. 1840.

KNIGHT, THOMAS ANDREW
POMONA HEREFORDIENSIS; CONTAINING COLOURED ENGRAVINGS OF THE OLD CIDER AND PERRY FRUITS OF HEREFORDSHIRE . . . ACCOMPANIED WITH A DESCRIPTIVE ACCOUNT OF EACH VARIETY.
4to (30½ cm × 24 cm). London 1811.
30 hand-coloured plates by E. Matthews and F. Knight, engraved by W. Hooker.

KNORR, GEORG WOLFGANG
THESAURUS REI HERBARIAE HORTENSISQUE UNIVERSALIS, EXHIBENS FIGURAS FLORUM, HERBARUM, ARBORUM, FRUTICUM, ALIARUMQUE, PLANTARUM PRORSUS NOVAS ET AD IPSOS DELINEATAS DEPICTASQUE ARCHETYPIS NATIVIS COLORIBUS . . . ALLGEMEINES BLUMEN-KRÄUTER-FRÜCHT-UND GARTENBUCH . . .
Title-page to vol. 3. REGNUM FLORAE. DAS REICH DER BLUMEN MIT ALLEN SEINEN SCHÖNHEITEN NACH DER NATUR UND IHREN FARBEN VORGESTELLT.
3 vols (2 vols and Atlas). Folio (39½ cm × 25 cm). Nürnberg (Nuremberg) 1770–72. According to Pritzel some copies of vol. 1 are dated 1750. c. 403 hand-coloured engraved plates and one undated hand-coloured engraved title page (to vol. 3) by G. W. Knorr. Text by P. F. Gmelin and G. R. Boehmer. Other editions Nuremberg 1779–82 and 1788.

KNOWLES, GEORGE BEAUCHAMP & WESTCOTT, FREDERICK
THE FLORAL CABINET AND MAGAZINE OF EXOTIC BOTANY.
3 vols 4to (26 cm × 20 cm). London 1837–1840.
137 hand-coloured lithographs. H.B.

KÖHLER, HERMANN ADOLPH
MEDIZINAL-PFLANZEN IN NATURGETREUEN ABBILDUNGEN . . . ATLAS SUR PHARMACOPOEA GERMANICA . . . (VOLS 1 AND 2) VON G. PABST . . . (VOL. 3) NEUESTE MEDIZINALPFLANZEN UND VERWECHSELUNGEN . . . VON M. VOGTHERR UND M. GÜRKE.
3 vols 4to (29 cm × 22 cm). Gera [1883–] 1887 [–98].
283 chromolithographed plates by L. Müller and C. F. Schmidt, lithographed by E. Günther.
Published in 70 parts.
From the botanical standpoint the finest and most useful series of illustrations of medicinal plants.

KOPS, JAN, & OTHERS
FLORA BATAVA, OF AFBEELDING EN BESCHRIJVING VAN NEDERLANDSCHE GEWASSEN . . .
28 vols 4to (30 cm × 24 cm). Amsterdam; Leiden (Leyden); Haarlem; The Hague 1800–1934.
c. 2240 coloured plates (vols 1–24 hand-coloured; vols 25–28 chromolithographed) by G. J. van Os, C. J. van Hulstijn, A. J. Kouwels, H. C. van de Pavord Smits.
Vols 1–10 (1800–49) edited by J. Kops, the rest by P.E.M. Gevers Deijnoot, F. A. Hartsen, F. W. van Eeden, L. Vuyck and W. J. Lütjeharms in succession.

KORTHALS, PIETER WILHELM
VERHANDELINGEN OVER DE NATUURLIJKE GESCHIEDENIS DER NEDERLANDSCHE OVERZEESCHE BEZITTINGEN, DOOR DE LEDEN DER NATUURKUNDIGE COMMISSIE IN INDIË EN ANDERE SCHRIJVERS.
Large Folio (43 cm × 29 cm). Leyden 1839–1842.
70 hand-coloured engravings. H.B.

KOTSCHY, CARL GEORG THEODOR
DIE EICHEN EUROPAS UND DES ORIENTS
Folio (53 cm × 35 cm). Wien [Vienna] and Olmütz [1858–] 1862.

LES CHÊNES DE L'EUROPE ET DE L'ORIENT. Paris 1864.
40 chromolithographed plates by Oberer and J. Seboth, lithographed by C. Horegschj.
The issues contain identical plates and text.

KOTSCHY, CARL GEORG THEODOR, & PEYRITSCH, JOHANN JOSEPH

PLANTES TINNÉENNES OU DESCRIPTIONS DE QUELQUES UNES DES PLANTES . . . DU BAHR-EL-GHASAL . . . EN AFRIQUE CENTRALE. PLANTAE TINNEANAE, SIVE DESCRIPTIO PLANTARUM.
Folio (55 cm × 39 cm). Vindobonae (Vienna) 1867.
27 hand-coloured plates (pls. 11–25, 8b, 15b), by Liepoldt and Mettenius, lithographed by Gebhard, Pauer, Polzer, Schoenhals, Strohmeyer and Wopalensky.
This book gives the botanical results of a tragic journey of exploration up the White Nile made in 1861–64 by three courageous Dutchwomen, Mme. Henriette Louise Marie Tinne, her daughter Alexandrie P. F. Tinne and her sister Adrienne van Capellen. The two elder women, Henriette and Adrienne, died on the journey.

KRAUSS, JOHAN CARL

AFBEELDINGEN DER FRAAISTE, MEEST UITHEEMSCHE BOOMEN EN HEESTERS. DIE TOT VERSIERING VAN ENGELSCHE BOSSCHEN EN TUINEN, OP ONZEN GROND, KUNNEN GEPLANT EN GEKWEEKT WORDEN . . .
21 parts 4to (29 cm × 23½ cm). Amsterdam 1802–08.
126 unsigned hand-coloured engraved plates, engraved by J. C. Sepp and Zoon.

KREBS, F. L.

VOLLSTÄNDIGE BESCHREIBUNG UND ABBILDUNG DER SÄMTLICHEN HOLZARTEN, WELCHE IM MITTLEREN UND NORDLICHEN DEUTSCHLAND WILD WACHSEN.
Vol. 1. 25 parts Folio (36½ cm × 26½ cm). Braunschweig 1826 [1827–35].
150 hand-coloured engraved plates (Nos 1–145, 28b, 43b, 44b, 45b, 124b) by A. Kohte, lithographed by Arckenhausen, Oehme and Müller, F. Sonntag, Strauber and Kohte.

KROCKER, ANTON JOHANN

FLORA SILESIACA RENOVATA EMENDATA CONTINENS PLANTAS SILESIAE INDIGENAS . . .
3 vols and supplement. 8vo (19 cm × 11 cm). Vratislaviae (Bratislava) 1787–1823.
111 unsigned hand-coloured engraved plates (Vol. 1, pls 1–52, 4b, 2 plates numbered 20, 24, 52, with pls 30, 34, 50 missing; Vol. 2, pls 1–44; Vol. 3, pls 1–9) engraved by J. D. Philippin, J. Berka, F. G. Endler.

KUNTH, CARL SIGISMUND

NOVA GENERA ET SPECIES PLANTARUM QUAS IN PEREGRINATIONE AD PLAGUM AEQUINOCTIALUM ORBIS NOVI COLLEGERUNT.
7 vols Folio (52 cm × 34 cm) and 4to (33½ cm × 25 cm). Paris 1815–25.
714 hand-coloured engraved plates (pls. 1–700, 332 bis, 481 bis, 483a, 499b, 514b, 532b, 547b, 547bis, 548bis, 562bis, 645bis, 647bis, 659bis, 660, 699bis) by Turpin, Humboldt, Poiteau, Kunth and L.C.M. Richard. This forms part of Humboldt and Bonpland's *Voyage aux Régions équinoxiales du Nouveau Continent . . . Botanique*. For dates of publication, see J. H. Barnhart in *Bull. Torrey Bot. Club* 29: 585–598 (1902), summarized in *Flora Malesiana* I. 4: cxc (1954). Often cited in botanical literature as 'H.B.K. Nov. Gen. Sp' although Kunth wrote the text; quarto and folio. Editions were issued simultaneously but differ in the pagination of the text as pointed out by Barnhart; botanists usually cite the quarto edition. Examination of parts 6–13 in wrappers as issued shows that Barnhart's deductions as to their contents were not wholly correct.
Vol. 2 (quarto) part 6, pp. 97–192, pls 122–146 . . . in Dec. 1817; part 7, pp. 193–280, pls 147–158, 162, 168, 172, 173 in Feb. 1818; part 8, pp. 281–406, pls 159–161, 163, 164–166, 170, 174–192 in June 1818.
Vol. 3 (quarto) part 9, pp. 1–96, half-title, titles, pls 193–217 (pl. 216 labelled *Aragoa juniperina*) in Sept. 1818; part 10, pp. 97–192, pls 218–242, in Feb. 1819; part II, pp. 193–288, pls 243–267, in July 1819; part 12, pp. 289–384, pls 268–292, in November 1819; part 13, pp. 385–456, pls 216 (labelled *Aragoa cupressina*) 293–300, in March 1820.

KUNTH, CARL SIGISMUND

MIMOSES, ET AUTRES PLANTES LÉGUMINEUSES DU NOUVEAU CONTINENT.
14 parts Folio (51½ cm × 34½ cm). Paris 1819 [–24].
60 hand-coloured plates by Turpin, engraved by Coignet, Dien, Joyau, Leroy, Plée fils, Teillard and Victor. This forms part of Humboldt and Bonpland's *Voyage . . . Botanique*. For evidence regarding dates of publication, see C. D. Sherborn and B. B. Woodward in *J. Bot. (London)* 39: 205 (1901) and reviews in *Naturwiss. Anzeiger* 4–5 (1820–22).
It may be summarised as follows:
Part 1, pp. 1–4, pls 1, 11, 36, 43, 50 . . . in June 1819;
part 2, pp. 5–16, pls 2–6 . . . in Sept. 1819;
part 3, pp. 17–28, pls, 8, 10, 19, 33, 34 . . . in Dec. 1819;
part 4, pp. 29–40, pls 7, 9, 12, 17, 18 . . . in Apr. 1820;
part 5, pp. 41–52, pls. 13–16, 20 . . . in July 1820;
part 6, pp. 53–72, pls 21, 22, 24, 27, 29 . . . in Dec. 1820;
part 7, pp. 73–84, pls 23, 24 (revised?), 25, 26, 27 (revised?), 28, 29 (revised?) . . . in June 1821;

part 8, pp. 85–96, pls 30, 31, 42, 45, 46 . . . in Nov. 1821;
part 9, pp. 97–108, pls 35–37, 39, 41 . . . In Jan. 1822;
part 10, pp. 109–120, pls. [32, 38, 40, 44, 47?] . . . in July 1822;
part 11, pp. 121–140, pls. [48, 49, 51?] . . . in July 1823;
part 12, pp. 141–160, pls [52–56?] . . . in Jan. 1824;
part 13, pp. 161–184, pls. [57–58?] . . . in May 1824;
part 14, pp. 185–223, pls. [59, 60?] . . . in June–July 1824.
The plate-contents of parts 10 to 14 are uncertain.

KUNTH, CARL SIGISMUND

REVISION DES GRAMINÉES PUBLIÉES DANS LES NOVA GENERA ET SPECIES PLANTARUM DE HUMBOLDT ET BONPLAND
2 vols Folio (52½ cm × 34 cm). Paris 1829–35.
220 hand-coloured engraved plates by E. Delile and Kunth. This is part of Humboldt and Bonpland's *Voyage . . . Botanique*. For dates of publication see C. D. Sherborn and B. B. Woodward in *J. Bot. (London)* 39: 205 (1901).

LA BILLARDIÈRE, JACQUES JULIEN HOUTON DE

ICONES PLANTARUM SYRIAE RARIORUM DESCRIPTIONIBUS ET OBSERVATIONIBUS ILLUSTRATAE.
5 parts 4to (26 cm × 20 cm). Lutetiae Parisiorum (Paris) 1791–1812.
50 uncoloured plates by Poiteau, P. J. and H. J. Redouté and Turpin, engraved by Aubry, Dien, Guyard, Maleuvre, Marchand, Milsan, Sellier and Voisar.

LA BILLARDIÈRE, JACQUES JULIEN HOUTON DE

NOVAE HOLLANDIAE PLANTARUM SPECIMEN.
2 vols 4to (34 cm × 25 cm). Parisiis (Paris) 1804–06 [-07].
265 uncoloured plates by La Billardière, Piron, Poiteau, P. J. Redouté, Sauvage, and Turpin, engraved by Dien, Plée, V. Plée fils, Sellier.
According to W. T. Stearn in *Flora Malesiana* I. 4: clxxi (1954), vol. 1, pls 1–10 published in 1804, pls 11–120 in 1805, pls 121–142 in 1806, vol. 2, pls 143–240 in 1806, pls 241–265 in 1807.

LA BILLARDIÈRE, JACQUES JULIEN HOUTON DE

SERTUM AUSTRO-CALEDONICUM.
4to (34½ cm × 25½ cm). Paris 1824–25.
80 engraved plates by Turpin.

LAGUNA Y VILLANUEVA, MAXIMO & AVILA, PEDRO DE

FLORA FORESTAL ESPAÑOLA, QUE COMPRENDE LA DESCRIPCIÓN DE LOS ÁRBOLES, ARBUSTOS Y MATAS, QUE SE CRÍAN SILVESTRES Ó ASILVESTRADOS EN ESPAÑA . . .
2 vols 8vo (26 cm × 18 cm) and Atlas, Folio (50 cm × 35½ cm). Madrid 1883–90.
80 chromolithographed plates by Justo de Salinas, chromolithographed by F. Aznar, M. Fuster, A. Gruas, T. Rufflé.

LAMARCK, JEAN BAPTISTE MONNET DE

[ENCYCLOPÉDIE MÉTHODIQUE] BOTANIQUE. RECUEIL DE PLANCHES DE BOTANIQUE DE L'ENCYCLOPÉDIE.
4 vols 4to. Paris 1789–1832.
With 1,000 uncoloured plates, of which 28 by Redouté.

LAMBERT, AYLMER BOURKE

A DESCRIPTION OF THE GENUS CINCHONA . . . ILLUSTRATED BY FIGURES OF ALL THE SPECIES HITHERTO DISCOVERED . . . ALSO A DESCRIPTION OF A NEW GENUS NAMED HYAENANCHE.
4to (26 cm × 21½ cm). London 1797.
13 folded plates, mostly (pls 4, 5, 7–12) by Ferdinand Bauer, engraved by J. Barlow.
Plate 1 is a copy of a plate in circulation in 1756.

LAMBERT, AYLMER BOURKE

A DESCRIPTION OF THE GENUS PINUS, ILLUSTRATED WITH FIGURES . . .
2 vols Folio (59 cm × 46½ cm). London 1803–24.
About 55 plates by Ferdinand and Franz Bauer, G. D. Ehret, S. Parkinson, J. Lindley, J. Sowerby, engraved by Barlow, Queiroz, Weddell, Warner and Mackenzie, usually uncolored, but 25 copies were coloured by W. Hooker. Vol. 1, 1803– [–07]. Vol. 2, to which is added an Appendix containing an account of the Lambertian Herbarium, by Mr Don, 1824.
For particulars, see H. W. Renkema and J. Ardagh in *J. Linn. Soc. Bot.* 48: 442–444 (1930). As noted there, 'again and again one finds citations of Lambert's *Pinus* which are contradictory and give rise to confusion. This is due to the fact that not only was the work issued over a long series of years but also and, to a high degree, to the circumstance that copies of what is apparently the same issue do not always agree in contents and arrangement. Each issue of the work appears to be made up, with many irregularities, from the material available at the time to the publishers'. From the bibliographical standpoint Lambert's *Pinus* is among the most exasperating of all botanical works since no two copies seem to be exactly the same. The plates are excellent.
This work is discussed in the text by Wilfrid Blunt.

A DESCRIPTION OF THE GENUS PINUS . . .
Ed. 2. To which is added an appendix containing descriptions and figures of some other remarkable plants, and an account of the Lambertian herbarium by David Don.
3 vols Folio (63 cm × 48 cm). London 1828–37.

About 80–89 plates by J. Sowerby and Franz Bauer, J. Lindley, J. Lycett, A. Manz, engraved by Barlow, W. Hooker, Mackenzie, Queroz, Warner, and Weddell and lithographed by Scharf.
For particulars see Renkema and Ardagh's paper, 444–47, 454–56 (1930).

A DESCRIPTION OF THE GENUS PINUS . . .
Ed. minor. 2 vols 8vo (37½ cm × 26½ cm). London 1832.
About 76 plates; Renkema and Ardagh, *loc. cit.*, 447 (1930) also mention copies with 72, 75, 76, 81, 103 and even 111 plates. 'Some of the plates are re-engravings of parts of, and a few are reductions from, the original ones . . ., some plates folded endlong and cut to octavo size.'

A DESCRIPTION OF THE GENUS PINUS . . .
Ed. 2. 3 vols Folio (64 cm × 47 cm). London 1837 [–42]
About 117 plates; see Renkema and Ardagh's paper, 448–49, 458–460 (1930).

A DESCRIPTION OF THE GENUS PINUS . . .
Ed. minor. 2 vols 8vo and Atlas Folio. London 1842.
As stated by Renkema and Ardagh, *loc. cit.* 449 (1930) 'the title pages of vols i and ii are identical with those of 1832 but with James Bohn as publisher and the date 'MDCCCXLII.' The publisher thought it desirable to issue a better edition in two volumes, consisting of the original plates in a separate volume and the text of the 1832 edition'.

LANGE, JOHAN MARTIN CHRISTIAN

DESCRIPTIO ICONIBUS ILLUSTRATA PLANTARUM NOVARUM VEL MINUS COGNITARUM PRAECIPUE E FLORA HISPANICA . . .
3 parts Folio (40 cm × 31½ cm). Havniae (Copenhagen) 1864 [–1866].
35 hand-coloured engraved plates by C. Thornam.

LAPEYROUSE, PHILIPPE PICOT DE

FIGURES DE LA FLORE DES PYRÉNÉES
Vol. I. 4 parts Folio. Paris. An III–IX (1795–1801).
43 plates in colour, of which 11 by Redouté.
Another imprint: An XI (1801), Toulouse.

LAVALLÉE, ALPHONSE

LES CLEMATIDES À GRANDES FLEURS. CLEMATIDES MEGALANTHES. DESCRIPTION ET ICONOGRAPHIE . . .
4to (34½ cm × 26½ cm). Paris 1884.
24 uncoloured plates (pls 1–22, 4bis, 7bis) by B. Bergeron and A. Riocreux, lithographed by Picart.
Published Jan. 1884.

LAVALLÉE, ALPHONSE

ARBORETUM SEGREZIANUM. ICONES SELECTAE ARBORUM ET FRUTICUM IN HORTIS SEGREZIANIS COLLECTORUM. DESCRIPTION ET FIGURES DES ESPÈCES NOUVELLES, RARES OU CRITIQUES DE L'ARBORETUM DE SEGREZ.
4to (35 cm × 26½ cm.). Paris [1880–] 1885.
36 uncoloured plates by B. Bergeron, C. Cuisin and A. Riocreux, lithographed by Picart. parts 1–2, pp. 1–40, pls 1–12 in 1880; parts 3, pp. 41–64, pls 13–18 in 1881; parts 4, pp. 65–88, pls 19–24 in 1882; parts 5, pp. 89–108, pls 25–30 in 1883; parts 6, pp. 109, pls 31–36 in 1885.

LAWRANCE, MARY

A COLLECTION OF PASSION FLOWERS.
Folio (46½ cm × 35 cm). London (1799–) 1802.
18 hand-coloured unnumbered etchings by M. Lawrance.
Discussed in text by Wilfrid Blunt.

LAWRANCE, MARY

A COLLECTION OF ROSES FROM NATURE.
30 parts Folio (47 cm × 37½ cm). London 1799 [1796–1810].
Frontispiece and 90 hand-coloured etchings by M. Lawrance.
Discussed in text by Wilfrid Blunt.

LEDEBOUR, CARL FRIEDRICH VON

ICONES PLANTARUM NOVARUM VEL IMPERFECTE COGNITARUM FLORAM ROSSICAM IMPRIMIS ALTAICUM ILLUSTRANTES.
5 vols Folio (46 cm × 31 cm). Rigae, Parisiis (Paris), Argentorati (Strasbourg), Londini (London) 1829–34.
500 hand-coloured plates by E. Bommer, W. Krüger, W. Müller, D. v. d. Pahlen, F. Scheffner, C. v. Ungern-Sternberg, lithographed by W. Siegrist, Prestele and Schach. For dates of publication, see W. T. Stearn in *J. Arnold Arb.* 22:228 (1941).

LEE, MRS SARAH (FORMERLY MRS BOWDICH)

TREES, PLANTS AND FLOWERS.
8vo. London 1854.
8 hand-coloured lithographs.
Mrs Bowdich was the authoress of the famous book on British fishes. H.B.

LEHMANN, JOHANN GEORG CHRISTIAN

MONOGRAPHIA GENERIS POTENTILLARUM.
4to (25 cm × 20½ cm). Hamburgi, Parisiis & Londini 1820.
20 plates, engraved and perhaps drawn by F. Guimpel.
Supplement parts 1. 4to (25 cm × 20½ cm). Hamburgi 1835.
10 unsigned plates, lithographed by Speckler & Co.

Pæonia Moutan. Var. b.

LEHMANN, JOHANN GEORG CHRISTIAN

ICONES ET DESCRIPTIONES NOVARUM ET MINUS COGNITARUM STIRPIUM (ICONES RARIORUM PLANTARUM E FAMILIA ASPERIFOLARUM).
5 parts Folio (37½ cm × 23 cm). Hamburg 1821 (–24).
50 uncoloured plates engraved (and drawn?) by J. F. Schröter.

LE MAOUT, EMMANUEL

BOTANIQUE. ORGANOGRAPHIE ET TAXONOMIE, HISTOIRE NATURELLE DES FAMILLES VÉGÉTALES.
Royal 8vo. Paris 1851–2.
50 plates (incl. 30 hand-coloured lithographs) H.B.

LE MAOUT, EMMANUEL. See also DECAISNE, JOSEPH 1858.

LEMAIRE, CHARLES. See Periodicals, JARDIN FLEURISTÈ, LE

L'HÉRITIER DE BRUTELLE, CHARLES LOUIS

STIRPES NOVAE, AUT MINUS COGNITAE, QUAS DESCRIPTIONIBUS ET ICONIBUS ILLUSTRAVIT.
6 parts Folio (56½ cm × 42½ cm). Parisiis (Paris) 1784–85 [i.e. 1785–91].
91 hand-coloured plates (pls 1–84, with two each numbered 7, 30, 52, 53, 56, 57, 59) by L. Freret, P. J. Redouté, J. C. Bruguière, Fossier, P. Jossigny, Prévost, J. Sowerby, engraved by Baron and others.
For dates of publication, see J. Britten and B. B. Woodward in *J. Bot. (London)* 43: 267 (1905). For list of extra unpublished plates, see W. T. Stearn in S. Sitwell and R. Madol, *Album de Redouté*, 19 (1954). It contains early work by P. J. Redouté.

L'HÉRITIER DE BRUTELLE, CHARLES LOUIS

GERANIOLOGIA, SEU ERODII, PELARGONII, GERANII, MONSONIAE ET GRIELI HISTORIA ICONIBUS ILLUSTRATA.
Folio (52 cm × 34 cm). Large paper ed. (60 cm × 45 cm). Parisiis (Paris) 1787–88 [i.e. 1791 or 1792].
44 engraved plates by P. J. Redouté, Aubriet, L. Freret, Pernotin, H. J. Redouté, J. Sowerby and Taylor, engraved by F. Hubert, Allix, Baron, L. Boutelon, J. B. Guyard, Juillet, Maleuvre, Milsan, S. Voysard.
No text issued.
Regarding date of publication, see J. Britten and B. B. Woodward in *J. Bot. (London)* 43: 271 (1905).

L'HÉRITIER DE BRUTELLE, CHARLES LOUIS

SERTUM ANGLICUM, SEU PLANTAE RARIORES QUAE IN HORTIS JUXTA LONDINUM, IMPRIMIS IN HORTO REGIO KEWENSI EXCOLUNTUR . . .
4 parts Folio (50½ cm × 34 cm). Parisiis (Paris) 1788 [–92].
35 engraved plates (Nos 1–34, 15 bis) by J. G. Bruguière, Pernotin, P. J. Redouté, J. Sowerby, engraved by J. B. Guyard, F. Hubert, Juillet, Maleuvre, Milsan, S. Voysard.
For dates of publication, see J. Britten and B. B. Woodward in *J. Bot. (London)* 43: 272 (1905).

L'HÉRITIER DE BRUTELLE, CHARLES LOUIS

CORNUS . . . DESCRIPTIONES ET ICONES SPECIERUM CORNI MINUS COGNITARUM.
Folio (57 cm × 43 cm). Paris 1788 [i.e. 1789].
6 hand-coloured engraved plates by L. Freret and P. J. Redouté, engraved by Devisse, F. Hubert, Juillet & Maleuvre.
According to J. Britten & B. B. Woodward in *J. Bot. (London)* 43: 272 (1905) this was published in 1789. The dates on L'Héritiers title-page are not to be trusted.

LIBOSCHITZ, JOSEPH, & TRINIUS, KARL BERNHARD

FLORE DES ENVIRONS DE ST PETERSBOURG ET DE MOSCOU.
1 vol. (all published). 4to (29 cm × 22½ cm). St. Pétersbourg (Leningrad) 1811.
41 hand-coloured engraved plates unsigned and unnumbered.

LIEBMANN, FREDERIK MICHAEL

LES CHÊNES DE L'AMÉRIQUE TROPICALE. ICONOGRAPHIE DES ESPÈCES NOUVELLES OU PEU CONNUES. OUVRAGE POSTHUME . . . ACHEVÉ ET AUGMENTE . . . PAR A. S. OERSTEDT.
Folio (45 cm × 34½ cm). Leipzig 1869.
10 nature prints (A–K) and 47 unsigned engraved plates by J. T. Bayer or C. Thornam, with analyses by Oerstedt.

LINDEN, JEAN JULES

PESCATOREA. ICONOGRAPHIE DES ORCHIDÉES . . . AVEC LA COLLABORATION DE J.E. PLANCHON, G. REICHENBACH FILS, G. LUDDEMANN.
1 vol. Folio (45 cm × 32 cm). Bruxelles (Brussels) (1854–) 1860.

PLATE 38

Aimé Bonpland wrote Description des plantes rares *(Paris 1813)*
to describe the luxurious gardens Empress Josephine maintained at Malmaison and Navarre.
Bouquet etched "Paeonia Moutan" in stipple
after the lifelike painting by Pierre Joseph Redouté.
The flower is inked in red, blue and green, and hand-colored
in a remarkable simulation of reality.

48 hand-coloured lithographed plates by F. Detollenaere and Maubert, lithographed by F. Detollenaere.

LINDLEY, JOHN

ROSARUM MONOGRAPHIA; OR A BOTANICAL HISTORY OF ROSES . . .
4to (24 cm × 15 cm). London 1820.
19 plates (18 hand-coloured) by J. Lindley and J. Curtis (pl. 9 only), engraved by S. Watts.

LINDLEY, JOHN

COLLECTANEA BOTANICA: OR FIGURES AND BOTANICAL ILLUSTRATIONS OF RARE AND CURIOUS EXOTIC PLANTS . . .
8 parts Folio (43½ cm × 30½ cm). London 1821 [–1825].
40 hand-coloured plates and 1 uncoloured by Ferd. Bauer, J. Curtis, M. Hart, W. Hooker, W. J. Hooker, B. Laurence, and J. Lindley, engraved by W. C. Edwards, C. Fox, S. Edwards and Weddell.
Plates 1–15 in 1821, pls 16–36 uncertain, pls 37–41 in 1825.

LINDLEY, JOHN

DIGITALIUM MONOGRAPHIA; SISTENS HISTORIAM BOTANICAM GENERIS, TABULIS OMNIUM SPECIERUM HACTENUS COGNITARUM . . .
Folio (46 cm × 32 cm). Londini (London). 1821.
28 hand-coloured engraved plates, 22 by Ferdinand Bauer, 1 by Franz Bauer and 5 by J. Lindley, engraved by Ferdinand Bauer, W. Hooker and J. Lindley.
An important work on foxgloves.

LINDLEY, JOHN

POMOLOGICAL MAGAZINE, THE; OR FIGURES AND DESCRIPTIONS OF THE MOST IMPORTANT VARIETIES OF FRUIT CULTIVATED IN GREAT BRITAIN.
3 vols 8vo (25½ cm × 16 cm). London 1828–30.
152 coloured engraved plates by C. M. Curtis and Mrs Withers, engraved by W. Clark and S. Watts.
Edited by John Lindley.
Reissued in 1841 as Lindley's *Pomologia Britannica.*

LINDLEY, JOHN

LADIES' BOTANY: OR A FAMILIAR INTRODUCTION TO THE STUDY OF THE NATURAL SYSTEM OF BOTANY.
2 vols 8vo (21½ cm × 14 cm). London 1834–37.
50 hand-coloured engraved plates by Miss Drake and F. Bauer, engraved by Watts.
Ed. 6. 1862.

LINDLEY, JOHN

VICTORIA REGIA.
Folio (73 cm × 52 cm). London 1837.
1 coloured lithographed plate by Schomburgk, lithographed by P. Gauci.
Only 25 copies issued, one of which Lindley presented to the Linnean Society of London on 7 Nov. 1837.

LINDLEY, JOHN

SERTUM ORCHIDACEUM: A WREATH OF THE MOST BEAUTIFUL ORCHIDACEOUS FLOWERS.
10 parts Folio (55½ cm × 37 cm). London (1837–)1838 (–41).
50 hand-coloured plates (Frontispiece and pls 1–49) by Miss Drake, Descourtilz, W. Griffith, M. A. Mearns, R. H. Schomburgk and Schouten, lithographed by M. Gauci. Plates dated 1837–1841.

LINDLEY, JOHN

POMOLOGIA BRITANNICA, OR FIGURES AND DESCRIPTIONS OF THE MOST IMPORTANT VARIETIES OF FRUIT CULTIVATED IN GREAT BRITAIN.
3 vols 8vo (23 cm × 13 cm). London 1841.
152 coloured plates by C. M. Curtis and Mrs Withers, engraved by W. Clark and S. Watts.
A reissue of *The Pomological Magazine* (1828–30) with a new title.

LINDLEY, JOHN, & PAXTON, SIR JOSEPH

PAXTON'S FLOWER GARDEN.
3 vols 4to (26 cm × 20 cm). London 1850–53.
108 hand-coloured lithographed plates by L. Constans and 314 uncoloured engraved text figures.
New Edition, revised by T. Baines.
3 vols 4to (27½ cm × 21½ cm). London. 1882–84.
107 chromolithographed plates by W. H. Fitch and 265 uncoloured engraved text figures.

LINK, HEINRICH FRIEDRICH & OTTO, CHRISTOPH FRIEDRICH

ICONES PLANTARUM SELECTARUM HORTI REGII BOTANICI BEROLINENSIS . . . ABBILDUNGEN AUSERLESENER GEWÄCHSE DES KÖNIGLICHEN BOTANISCHEN GARTENS ZU BERLIN . . .
1 vol. (10 parts). 4to (26½ cm × 21½ cm). Berlin 1820–28 [–29].
60 hand-coloured plates by F. Guimpel and C. Röthig, engraved by F. Guimpel and M. and P. Haas.
For dates of publication, see W. T. Stearn in *J. Soc. Bibl. Nat. Hist.* 1:105 (1937).

LINK, HEINRICH FRIEDRICH & OTTO, CHRISTOPH FRIEDRICH

ICONES PLANTARUM RARIORUM HORTI REGII BOTANICI BEROLINENSIS . . . ABBILDUNGEN NEUER UND SELTENER GEWÄCHSE DES KONIGLICHEN BOTANISCHEN GARTENS ZU BERLIN . . .
1 vol. (8 parts). 4to (21 cm × 15½ cm). Berlin 1828 [–31].

48 plates by F. Guimpel and C. Röthig, engraved by F. Guimpel and Weidlich.
For dates of publication, see W. T. Stearn in *J. Soc. Bibl. Nat. Hist.* I: 106 (1937).

LINK, HEINRICH FRIEDRICH; KLOTZSCH, JOHANN FRIEDRICH, & OTTO, CHRISTOPH FRIEDRICH

ICONES PLANTARUM RARIORUM REGII BOTANICI BEROLINENSIS. ABBILDUNGEN SELTENER PFLANZEN DES KÖNIGL. BOTANISCHEN GARTENS ZU BERLIN.
2 vols 4to (27 cm × 22 cm). Berlin [1840–] 1841–44.
48 chromolithographed plates by C. F. Schmidt.
For dates of publication, see W. T. Stearn in *J. Soc. Bibl. Nat. Hist.* 1: 106 (1937).

LINNAEUS, CARL, later CARL VON LINNÉ

HORTUS CLIFFORTIANUS, PLANTAS EXHIBENS QUAS IN HORTUS TAM VIVIS QUAM SICCIS HARTECAMPI IN HOLLANDEA COLUIT GEORGIUS CLIFFORD.
Folio (41 cm × 25 cm). Amstelaedami (Amsterdam) 1737.
Frontispiece and 36 uncoloured plates by G. D. Ehret and J. Wandelaar, engraved by J. Wandelaar.
A work of great botanical importance, in which Linnaeus gave concise definitions and elaborate synonyms for the numerous species grown in the garden of George Clifford at Hortekamp or represented in his herbarium. When Linnaeus dealt later with the same species in his Species Plantarum (1753) *he cited the* Hortus Cliffortianus *whenever possible as a source of further information and it thus has an important bearing on the application of Linnaean botanical names.*

LODDIGES, CONRAD. See Periodicals.

LOISELEUR DESLONGCHAMPS, JEAN LOUIS AUGUSTE

FLORA GALLICA, SEU ENUMERATIO PLANTARUM IN GALLIA SPONTE NASCENTIUM.
2 vols 8vo (16 cm × 9½ cm). Lutetiae (Paris) 1806–07.
21 engraved plates, all folded, by A. L. Marquis.
Ed. 2. Aucta et emendata.
2 vols 8vo (20½ cm × 13 cm). Parisiis (Paris) 1828.
31 engraved plates by Logerot and A. L. Marquis, engraved by A. L. Marquis and A. Tardieu, pl. 1–21 as in ed. 1, pl. 22–31 additional.

LOISELEUR DESLONGCHAMPS, JEAN LOUIS AUGUSTE

FLORE GÉNÉRALE DE FRANCE, OU ICONOGRAPHIE, DESCRIPTION ET HISTOIRE DE TOUTES LES PLANTES . . . QUI CROISSENT DANS CE ROYAUME PHANÉROGAMIE . . .
1 vol. 8 parts 8vo (22 cm × 14 cm). Paris 1828 [–1832].
96 plates (Nos I–XII uncoloured, and Nos 1–84 hand-coloured) by Poiteau, engraved by P. Duménil.

LOUDON, MRS [JANE WEBB]

THE LADIES' FLOWER GARDEN OF ORNAMENTAL ANNUALS.
4to (29 cm × 22 cm). London 1840.
48 hand-coloured plates by Mrs Loudon, lithographed by Day and Haghe.
2nd edition n.d. [1850]. H.B.

LOUDON, MRS [JANE WEBB]

THE LADIES' FLOWER GARDEN OF ORNAMENTAL BULBOUS PLANTS.
4to (29 cm × 22 cm). London 1841.
58 unsigned hand-coloured lithographed plates (containing 305 figs.) by Mrs Loudon, lithographed by Day and Haghe.
2nd edition n.d. [1850]. H.B.

LOUDON, MRS [JANE WEBB]

THE LADIES' FLOWER GARDEN OF ORNAMENTAL PERENNIALS.
2 vols 4to (29 cm × 22 cm). London 1843–44.
96 unsigned hand-coloured lithographs by Mrs Loudon, lithographed by Day and Haghe.
2nd edition n.d. (1850) with 90 hand-coloured lithographs.
The second edition of this and all Mrs Loudon's other flower books is very much inferior in the quality of the plates. H.B.

LOUDON, MRS [JANE WEBB]

THE LADIES' FLOWER GARDEN OF ORNAMENTAL GREENHOUSE PLANTS.
4to (29 cm × 22 cm). London 1848.
42 unsigned hand-coloured lithographs by Mrs Loudon.
2nd edition n.d. [1850]. H.B.

LOUDON, MRS [JANE WEBB]

BRITISH WILD FLOWERS.
4to (29 cm × 22 cm). London 1846.
60 hand-coloured lithographs.
2nd edition 1849.
3rd edition n.d. [1850]. H.B.

MACDONALD, ALEXANDER (Pseudonym)

A COMPLETE DICTIONARY OF PRACTICAL GARDENING.
2 vols 4to (30 cm × 23 cm). London 1807.
61 hand-coloured engravings from drawings by Sydenham Edwards (q.v.). H.B.

McINTOSH, CHARLES

FLORA AND POMONA; OR THE BRITISH FRUIT AND FLOWER GARDEN; CONTAINING DESCRIPTIONS OF THE MOST VALUABLE AND INTERESTING FLOWERS AND FRUITS, CULTIVATED IN THE GARDENS OF GREAT BRITAIN, WITH FIGURES DRAWN AND COLOURED FROM NATURE . . .

Pl. 50.

Antirrhinum latifolium. *Muflier à larges feuilles.*

Peint par G.W. Völker. Gravé par F.W. Meyer.

4to (28 cm × 22 cm). London 1829 [-32].
71 hand-coloured engraved plates (Nos 1–35, 1–36) by E. D. Smith, engraved by S. Watts.

McINTOSH, CHARLES
THE FLOWER GARDEN.
8vo (17 cm × 11 cm). London 1838.
Frontispiece, title and 9 plates, all hand-coloured lithographs.
2nd edition 1839. H.B.

McINTOSH, CHARLES
THE GREENHOUSE.
8vo (17 cm × 11 cm). London 1838.
Coloured title-page and 17 hand-coloured lithographs. H.B.

McINTOSH, CHARLES
THE ORCHARD.
8vo (17 cm × 11 cm). London 1839.
Coloured title-page and 17 hand-coloured lithographs. H.B.

MAKINO, TOMITARO
ILLUSTRATIONS OF THE FLORA OF JAPAN TO SERVE AS AN ATLAS TO THE NIPPON SHOKUBUTSUSHI.
4to (27 cm × 18½ cm). Tokyo (1887–) 1890 [-91].
72 lithographed plates (Nos 1–69, 37b, 38b, 40b) by T. Makino.

MALO, CHARLES
GUIRLANDE DE FLORE.
12mo (12 cm × 8 cm). Paris n.d. but probably 1815.
15 hand-coloured engraved plates and a title-page in colour by P. Bessa.
Exquisite miniature paintings of popular florists' flowers of the period. P.M.S.

MARSCHALL VON BIEBERSTEIN, FRIEDRICH AUGUST
CENTURIA PLANTARUM RARIORUM ROSSIAE MERIDIONALIS PRESERTIM TAURIAE ET CAUCASI, ICONIBUS DESCRIPTIONIBUSQUE ILLUSTRATA.
2 parts Folio (57 cm × 34 cm). Part I, Charkoviae (Charkov). 1810.
Part 2, Petropoli (St. Petersburg) 1832–43.
80 hand-coloured unsigned engraved plates.

MARTIUS, KARL FRIEDRICH PHILIPP VON
HISTORIA NATURALIS PALMARUM.
3 vols. I. De palmis generatim (with H. A. Mohl and F. Unger). 2. De Brasiliae palmis singulatim. 3. Expositio palmarum systematia.
Folio (62 cm × 46 cm). Monachii (Munich), Lipsiae (Leipzig) 1823–50.
240 hand-coloured lithographed plates (Anat. and Morph. A–Z, excluding J, Z1–Z23, tabs. geol. 1–3; Nos 1–180, 6a, 11a, 18a, 50a, 59a, 73a–d, 77a) by Martius, Hellmuth, Prestele, Minsinger, lithographed by Ezdorf, C. Hess, F. Hohe, Ziegler, A. Falger, Hellmuth and others.
Parts 1–9 issued as *Genera et Species Palmarum*. For dates of publication, see B. B. Woodward in *J. Bot. (London)* 46: 197 (1908), M. J. van Steenis Kruseman and W. T. Stearn in *Flora Malesiana* I. 4. cxcix (1954).
The finest book on Palms ever issued.

MARTIUS, KARL FRIEDRICH PHILIPP VON
NOVA GENERA ET SPECIES PLANTARUM QUAS IN ITINERE PER BRASILIAM ANNIS 1817–20 JUSSU ET AUSPICIIS MAXIMILIANI JOSEPHI I . . . SUSCEPTO COLLEGIT ET DESCRIPSIT C. F. P. MARTIUS . . . PINGENDAS J.G. ZUCCARINI.
3 vols Folio (36 cm × 27 cm). Monachii (Munich) [1823–] 1824–29 [-32].
300 hand-coloured lithographed plates by J. G. Zuccarini, T. Bischoff, Hellmuth, A. Manz, S. Minsinger, J. Prestele, lithographed by J. Prestele, W. Siegrist, A. Falger, Päringer, coloured by S. Minsinger. Title-page by Neureuther, engraved by Prestele.
Title-page bearing title as above appears in vols 1, 2 and 3 except that Zuccarini's name is not on those for vols 2 and 3.
Each volume also contains the decorative title-page of Neureuther with the title 'Nova Genera et Species Plantarum Brasiliensium.'
Vol. 1 has also a third title-page, as follows:
Nova Genera et Species plantarum quas in itinere annis 1817–1830 per Brasiliam jussu et auspiciis Max. Jos. 1 . . . collegit et descripsit C.F.P. Martius.
Opus tria volumina continens cum tabulis ccc. Monachii. 1823–32.
For dates of publication see B. B. Woodward in *J. Bot. (London)* 46: 197 (1908).

MARTIUS, KARL FRIEDRICH PHILIPP VON
[AMOENITATES BOTANICAE MONACENSES] AUSWAHL

PLATE 39
Color-printed stipple etching was not a monopoly of the French, as proved by this striking image. "Antirrhinum latifolium" illustrates Johann Hoffmannsegg's Flore portugaise *(Berlin 1809–40). Artist G. W. Voelcker has recreated a living plant, and etcher F. W. Meyer has conveyed it masterfully.*

MERKWÜRDIGER PFLANZEN DES K. BOTANISCHEN GARTENS ZU MÜNCHEN . . . CHOIX DES PLANTES REMARQUABLES . . .
4 parts 4to (27 cm × 21 cm). Frankfurt am Main [1829–31]. Paris 1831.
16 unsigned hand-coloured lithographed plates.
The Latin title appeared only on the wrappers. For dates of publication, see B. B Woodward in *J. Bot. (London)* 46: 198 (1908).

MARTIUS, KARL FRIEDRICH PHILIPP VON & OTHERS

FLORA BRASILIENSIS. ENUMERATIO PLANTARUM IN PORASILIA HACTENUS DETECTARUM, QUAS QUIS ALIORUMQUE BOTANICIS STUDIIS DESCRIPTAS ET METHODO NATURALI DIGESTAS PARTIM ICONE ILLUSTRATAS EDITIT C. F. P. DE MARTIUS (LATER A.G. EICHLER AND I. URBAN).
15 vols Folio (46 cm × 30 cm). Monachii (Munich); Vindobonae (Vienna); Lipsiae (Leipzig) 1840–1906.
3811 lithographed plates and nature-prints by Martius, Seboth and many others, lithographed by E. Bolmann, A. Brandmeyer, F. Hohe, M. Kuher, C. A. Lebsche, Keller, Minsinger, Obpacher Prestche, Prillwitz, Stoppel and others.

MARTYN, JOHN

HISTORIA PLANTARUM RARIORUM.
5 parts. Folio (52 cm × 35 cm). Londini (London) 1728 (–32).
50 unnumbered colour-printed plates by J. van Huysum, W. Houstoun, Massey, R. Sartorius, G. Sartorys, engraved by E. Kirkall.
Stated by Pulteney in 1790 to be 'the most sumptuous and magnificent work of the kind, that had ever been attempted in England'. Its plates, with those of the *Catalogus Plantarum* (1730) published by the Society of Gardeners, are the earliest examples of colour-printing in botanical books.

MARTYN, JOHN

HISTORIA PLANTARUM RARIORUM . . . DENUO EDITA STUDIO ATQUE OPERA JOANNIS DANIELIS MEYERI PICTORIA. JOHANN MARTYNS BESCHREIBUNG SELTENER PFLANZEN . . .
Folio (45 cm × 30 cm). Nürnberg (Nuremberg) 1752.
50 partly hand-coloured, partly colour-printed engraved plates copied by J. D. Meyer from the London edition. Martyn's Latin text is reprinted with a German translation. A similar later edition.

ABBILDUNG UND BESCHREIBUNG SELTENER GEWÄCHSE, NEU ÜBERSETZT VON GEORG WOLFGANG FRANZ PANZER . . .
Folio (47 cm × 33 cm). Nürnberg (Nuremberg), 1797.
50 coloured engraved plates by J. D. Meyer as above but with a new German translation of the Latin text.

MARTYN, THOMAS

THIRTY-EIGHT PLATES WITH EXPLANATIONS; INTENDED TO ILLUSTRATE LINNAEUS'S SYSTEM OF VEGETABLES, AND PARTICULARLY ADAPTED TO THE LETTERS ON THE ELEMENTS OF BOTANY [ADDRESSED TO A LADY BY THE CELEBRATED JEAN JACQUES ROUSSEAU].
8vo (21½ cm × 13 cm). London 1788.
38 hand-coloured engraved plates by F. P. Nodder.
2nd Edn. 1794.

MARTYN, THOMAS

FLORA RUSTICA, EXHIBITING ACCURATE FIGURES OF SUCH PLANTS AS ARE EITHER USEFUL OR INJURIOUS IN HUSBANDRY . . . WITH SCIENTIFIC CHARACTERS, POPULAR DESCRIPTIONS AND USEFUL OBSERVATIONS.
4 vols 8vo (21½ cm × 13 cm). London 1792–94.
144 hand-coloured engraved plates by F. P. Nodder, coloured under his inspection.

MASSON, FRANCIS

STAPELIAE NOVAE: OR A COLLECTION OF SEVERAL NEW SPECIES . . . DISCOVERED IN . . . AFRICA.
Folio (37½ cm × 26 cm). London 1796 (–97).
41 hand-coloured plats by F. Masson, engraved by Mackenzie.
Drawings made by Masson in S. Africa.

MAUND, BENJAMIN. See Periodicals.

MAW, GEORGE

A MONOGRAPH OF THE GENUS CROCUS.
4to (31 cm × 25 cm). London. 1886.
71 hand-coloured plates (A–D, 1–67) by G. Maw, lithographed by F. Huth.
Maw's drawings 'most exquisite' said Ruskin. The most important work on the genus. The plates, though rather stiff and wooden, are wonderfully detailed and accurate and there are numerous vignetted steel engravings of places in Asia Minor and elsewhere near the habitats of crocuses.

MAY, A. E. & MAY, W.

CHOICE FLOWERS, A COLLECTION OF FAVOURITE DRAWINGS FROM THE GARDEN AND THE CONSERVATORY.
Folio (36 cm × 26 cm). London, Ackermann, 1849.
Drawn, lithographed and coloured from Nature by A. E. May, with descriptive letterpress by W. May, F.H.S.
Coloured frontispiece and 31 hand-coloured lithographs.
75 people only subscribed to this book. H.B.

MEEN, MARGARET

EXOTIC PLANTS FROM THE ROYAL GARDENS AT KEW.
3 parts folio (61 cm × 47 cm). London 1790–95.
9 aquatint and colour-printed plates by M. Meen, engraved by M. C. Prestel.
An indefatigable gifted amateur. W.B.

MEERBURG (MEERBURGH), NICOLAAS

AFBEELDINGEN VAN ZELDZAAME GEWASSEN.
Folio (42½ cm × 26 cm). Leyden 1775 [–1782?].
50 hand-coloured engraved plates by N. Meerburg.

MEERBURG (MEERBURGH), NICOLAAS

PLANTAE RARIORES VIVIS COLORIBUS DEPICTAE.
Folio (44 cm × 27 cm). Lugduni Batavorum (Leyden) 1789.
55 hand-coloured engraved plates by N. Meerburg.

MEERBURG (MEERBURGH), NICOLAAS

PLANTARUM SELECTARUM ICONES PICTAE.
Folio (40 cm × 25 cm). Lugduni Bat. (Leyden) 1798.
28 hand-coloured engraved plates by N. Meerburg.

MERIAN, MARIA SIBYLLA

METAMORPHOSIS INSECTORUM SURINAMENSIUM. OFTE VERANDERING DER SURINAAMSCHE INSECTEN.
Folio (52 cm × 36 cm). Amsterdam 1705.
60 hand-coloured plates by M. S. Merian, engraved by J. Mulder, J. P. Sluyter and D. Stoopendaal.

DISSERTATIO DE GENERATIONE ET METAMORPHOSIBUS INSECTORUM SURINAMENSIUM . . . AD VIVUM ACCURATE DEPICTA . . . M.S. MERIAEN OVER DE VOORTTEELING EN WONDERBAERLYKE VERANDERINGEN DER SURINAEMSCHE INSECTEN . . .
Folio (52 cm × 36 cm). Amsteldami (Amsterdam). 1719.
Frontispiece and 72 plates by M. S. Merian, engraved by J. Mulder, J. P. Sluyter and D. Stoopendaal.
Title-page of 1719 is in Latin only.
Frontispiece is by F. Ottens, engraved by J. Oosterwijk.
Discussed by Wilfrid Blunt in text.

MERIAN, MARIA SIBYLLA

HISTOIRE GÉNÉRALE DES INSECTES DE SURINAM ET DE TOUT L'EUROPE TROISIÈME ÉDITION, REVUE, CORRIGÉE ET CONSIDÉRABLEMENT AUGMENTÉE PAR M. BUCHOZ.
3 vols Large folio (54 cm × 30 cm). Paris 1771.
Vol. 1 hand-coloured title-pages and 72 hand-coloured engravings.
Vol. 2 182 hand-coloured engravings, 3 or 4 on each page.
Vol. 3 69 hand-coloured engravings. H.B.

MEYER, GEORG FRIEDRICH WILHELM

FLORA DES KÖNIGREICHS HANNOVER, ODER SCHILDERUNG SEINER VEGETATION . . .
Folio (54 cm × 37½ cm). Göttingen 1842–54.
Title-page and 30 partly colour-printed, partly hand-coloured engraved plates by Eberlein, F. W. Meyer, Saxesen, Schumann, Sontag, M. Tettelbach, engraved by Andorff, F. Fleischmann, Grape, C. Guinand, F. W. Meyer, C. Meyer, H. Meyer, C. Schmelz, Schumann.

MICHAUX, ANDRÉ

HISTOIRE DES CHENES DES L'AMÉRIQUE, OU DESCRIPTIONS ET FIGURES DE TOUTES LES ESPÈCES ET VARIÉTÉS.
Folio (45 cm × 31 cm). Paris An IX (1801).
36 plates by P. J. & H. J. Redouté, engraved by Plée & Sellier.

MICHAUX, ANDRÉ

FLORA BOREALI-AMERICANA . . .
2 vols 8 vo (23 cm × 15 cm). Parisiis & Argentorati (Paris & Strasbourg) An. XI (1803).
51 uncoloured plates by P. J. Redouté, engraved by Plée.
On sale in March 1803, according to B. G. Schubert in *Rhodora* 44: 149 (1942).
Second impression. Parisiis (Paris) 1820.

MICHAUX, FRANÇOIS ANDRÉ

HISTOIRE DES ARBRES FORESTIERS DE L'AMÉRIQUE SEPTENTRIONALE . . . LEURS USAGE DANS LES ARTS ET . . . DANS LE COMMERCE.
3 vols 4to (26 cm × 16½ cm). Paris 1810–13.
138 coloured plates (vol. 1, pls 1–14, 1–10; vol. 2, pls 1–26, 1–24; vol. 3, pls 1–11, 1–13, 1–11, 1–11, 1–13, 1–5) by Bessa, P. J. and H. J. Redouté, and A. Riché, engraved by Bessin, Boquet, Cally, Dubreuil, Gabriel, L. F. Jacquinot, J. N. Joly and Renard.
Issued as follows:
Vol. 1. parts 1–2, pls 'Pins' 1–14 . . . 1810. (Issued separately as *Histoire des Pins et Sapins*); parts 3–4, pls 'Noyers' 1–10. . . 1811 (issued separately as *Histoire des Noyers*);
Vol. 2. parts 5–8, pls 'Chenes' 1–26 . . . 1811 (issued separately as *Histoire des Chênes*); parts 9–10, pls 'Bouleaux' 1–12 . . . 1811; parts 11–12, pls 'Bouleaux' 13–24 . . . 1812.
Vol. 3. parts 13–14, pls 1–11 in 1812; parts 15–16, pls 1–13 in 1812; parts 17–18, pls 1–11 in 1812; parts 19–20, pls 1–11 in 1812; parts 21–22, pls 1–13 in 1813; parts 23–24, pls 1–5 in 1813.
See W. T. Stearn in Sitwell and Madol, *Album de Redouté* 22. (1954).

MIÇHAUX, FRANÇOIS ANDRÉ

THE NORTH AMERICAN SYLVA, OR A DESCRIPTION OF THE FOREST TREES OF THE UNITED STATES, CANADA AND NOVA SCOTIA . . . TO WHICH IS ADDED A DESCRIPTION OF . . . EUROPEAN FOREST TREES . . .

3 vols 8vo (25½ cm × 16½ cm). Philadelphia 1817–19.

156 hand-coloured plates by Bessa, P. J. and H. J. Redouté and A. Riche, engraved by Bessin, Boquet, Cally, Dubreuil, Gabriel, L. F. Jacquinot, J. N. Joly, Renard.

TRANSLATED FROM THE FRENCH . . .

3 vols 8vo (26 cm × 22 cm). Philadelphia 1857. Revised, 1859 and 1865.

156 hand-coloured engraved plates as above, printed from the original French copper-plates, with notes by J. Jay Smith.

MIKAN, JOHANN CHRISTIAN

DELECTUS FLORAE ET FAUNAE BRASILIENSIS . . .

Folio (55 cm × 40½ cm). Vindobonae (Vienna) 1820 [–25].

25 hand-coloured unnumbered lithographed plates (of which 12 are botanical) by Ferdinand Bauer (pls 1, 7 and analyses of 2), Buchberger (pls 2, 3, 8, 9, 14), Satory (19, 21), Stoll (pl. 20), Brunner, Funk and Sandler, lithographed by Erminy, Knapp, Kolb and Satory. Part 1 containing plates of *Stifftia chrysantha, Conchocarpus macrophyllus, Dichorisandra thysiflora*, published late in 1820; part 2 with plates of *Vellosia candida, Esterbazya splendida, Oxalis rusciformis*, in mid 1822; part 3, with plates of *Metternichia principis, Echites tenuifolia, Griffinia tenuifolia*, in July–Oct. 1823; part 4 with plates of *Helicteres brasiliensis, Passiflora amethystina, Gloxinia Schottii* in 1825; see A. Wetmore in *The Auk* 42. 283 (1925). W. T. Stearn in *J. Soc. Bibl. Nat. Hist.* 3: 135 (1956). The work was never completed and has no index or table of contents.

MILLER, JOHN (formerly JOHANN SEBASTIAN MÜLLER)

ILLUSTRATIO SYSTEMATIS SEXUALIS LINNAEI. AN ILLUSTRATION OF THE SEXUAL SYSTEM OF THE GENERA PLANTARUM OF LINNAEUS . . .

20 parts Folio (52½ cm × 35½ cm). London (1770–) 1777.

108 hand-coloured engraved plates (pls. 1–4 of leafform numbered, the other 104 pls. unnumbered without lettering and dates) and a duplicate set uncoloured, with lettering and dates, by J. Miller. For a note on this work, see W. B. Hemsley in *Gard. Chron.* III. 7: 255 (1890). It was issued in 20 parts from 1775 to 1777 but isolated plates may have been issued as printed from 1770 onwards: certainly Miller sent some to Linneaus in October 1772 and in 1773 and 1775. Greatly impressed, Linnaeus wrote that Miller's plates were more beautiful and accurate than any seen since the beginning of the world.

Pl. 29 is important botanically as the place of first publication of the description of *Cassyta baccifera* (= *Rhipsalis baccifera*).

New issue 1794.

The words 'of the genera plantarum' are omitted from the title-page of this issue. The coloured plates have lettering and plates.

MILLER, JOHN (formerly JOHANN SEBASTIAN MÜLLER)

AN ILLUSTRATION OF THE SEXUAL SYSTEM OF LINNAEUS. VOL. I

8vo (23 cm × 14 cm). London 1779.

105 coloured engraved plates ((i), 1–106 with 101 & 104 missing) and frontispiece by J. Miller. New issue 1794.

TABULAE ICONUM CENTUM QUATUOR PLANTARUM AD ILLUSTRATIONEM SYSTEMATIS SEXUALIS LINNAEANI . . . NOMINA PLANTARUM . . . A D.F.G. WEISS.

Vol. 2 8vo (21½ cm × 15 cm). Francofurti ad Moenum (Frankfurt am Main) 1789.

105 engraved plates (Tab. A, pls 1–104) by J. Miller, engraved by C. Goepfert.

These plates are copied from the 8vo *An illustration of the sexual System*.

AN ILLUSTRATION OF THE TERMINI BOTANICI OF LINNAEUS. VOL. II

8vo (23 cm × 14 cm). Lambeth (London) 1789.

85 coloured engraved plates (1–86, No 20 missing) by J. Miller.

MILLER, JOHN (formerly JOHANN SEBASTIAN MÜLLER)

[ICONES NOVAE].

Folio (50 cm × 35 cm). London 1780.

7 hand-coloured engraved plates by J. Miller. No title-page issued. Copies at British Museum (Natural History), the Royal Botanic Gardens, Kew, and the Lindley Library, Royal Horticultural Society. Plants figured are *Sophora tetraptera, Phormium tenax, Stewartia Malacodendron, Fothergilla latifolia, Heliconia Bihai, Lagerstroemia indica*.

For a note on this rare work see T. A. Sprague in *J. Bot (London)* 74: 208 (1936).

MILLER, PHILIP

THE GARDENER'S DICTIONARY . . .

Small folio. London 1731.

1st ed. engraved frontispiece and several text figs. of buildings etc.

2nd ed. 1733. Appendix 1735.

3rd ed. 1737.

4th ed. 1743.

Second [additional] volume 1739, 2nd ed. 1740.
6th ed., the former two volumes arranged in one alphabet, new engraved frontispiece by S. Wale.
7th ed. 1759 arranged according to Linnean System. Engraved frontispiece as in 6th ed. and 14 uncoloured engraved plates by J. Miller.
8th ed. 1768, revised. Plates as in 7th ed.
New edition entitled THE GARDENER'S AND BOTANIST'S DICTIONARY corrected and newly arranged by T. Martyn.
1807 2 vols each issued in 2 parts.
Vol. I, Part I with 14 engraved uncoloured plates unsigned.
Vol. I, Part 2 with 5 engraved uncoloured plates unsigned, two of which appeared in Part I.
Vol. II, Parts I and 2 without plates.
A most valuable and practical work, probably used widely over nearly 100 years and the forerunner of later Gardening Dictionaries. Miller had a number of collaborators in compiling it.

MILLER, PHILIP
FIGURES OF THE MOST BEAUTIFUL . . . PLANTS DESCRIBED IN THE GARDENER'S DICTIONARY . . .
2 vols Folio (41 cm × 26 cm). London (1755–) 1760.
300 hand-coloured or uncoloured engraved plates by G. D. Ehret, R. Lancake, J. S. Miller, W. Houstoun, J. Bartram, engraved by J. S. Miller, T. Jefferys, J. Mynde.
Issued in 50 monthly parts, the plates dated from 25 March 1755 to 30 July 1760; 47 parts published by 31 March 1759 (according to last page of Miller, Gard. Dict. Ed. 7).
Other editions 1771 and 1809.

MILLER, PHILIP, See also GARDENERS, SOCIETY OF 1730

MINER, HARRIET STEWART
ORCHIDS: THE ROYAL FAMILY OF PLANTS.
Folio. London 1885.
25 chromolithographs. H.B.

MIQUEL, FRIEDRICH ANTON WILHELM
CHOIX DE PLANTES . . . CULTIVÉES ET DESSINÉES DANS LE JARDIN BOTANIQUE DE BUITENZORG.
Folio (54 cm × 36 cm). La Haye (The Hague) 1864.
26 chromolithographed plates by J. Ljung and T. Rocke, lithographed by C. W. Mieling.

MOGGRIDGE, JOHN TRAHERNE
CONTRIBUTIONS TO THE FLORA OF MENTONE . . .
4 parts 8vo (25 cm × 15½ cm). London [1864–] 1871.
99 hand-coloured plates (pls. 1–97, 51 bis, 52 bis) by J. T. Moggridge, lithographed by V. Brooks and W. Dickes.
3rd Edition. 1874.
Part 1, pp. i–vii, pls 1–25 in 1864 (ed. 2, 1867; ed. 3, 1874); part 2, pls 26–50 in 1865; part 3, pls 51–73 in 1868; part 4, pls 74–97 in 1871.
The editions differ in the nomenclature of the plants figured, in part 1, e.g., pl. 1. *Anemone pavonina* (1864); *A. hortensis B. fulgens* and *Y. pavonina* (1867); *A. hortensis B ocellata* (1874); pl. 2. *Anemone stellata* (1864); *A. hortensis & A. stellata* (1867, 1874); pl. 12. *Primula latifolia* (1864, 1867); *P. viscosa* (1874); pl. 18. *Orchis olbiensis ?* (1864); *O. olbiensis* Reuter (1867, 1874); pl. 19. *Orchis Scolopax* (1864); *O. insectifera* (1867); *O. insectifera* subsp. *arachnites* var. *pseudo-scolopax* (1874); pl. 21. *Leucoium biemale* (1864); *L. biemale var.* (1867); *L. nicaëense* (1874); pl. 25. *Frittillaria delphinensis* Grenier ? (1864, 1867); *F. delphinensis* var *Y Moggridgei* (1874).
Moggridge was a beautiful draughtsman. The lithography does little justice to the original drawings. W.B.

MORIARTY, HENRIETTE MARIA
VIRIDARIUM: COLOURED PLATES OF GREENHOUSE PLANTS, WITH THE LINNEAN NAMES, AND WITH CONCISE RULES FOR THEIR CULTURE.
8vo (19 cm × 11 cm). London 1806.
50 hand-coloured engravings.
2nd edition 1807. H.B.

MORICAND, STEFANO
PLANTES NOUVELLES D'AMÉRIQUE.
10 parts Folio (32 cm × 25 cm). Genève (Geneva) 1833–46.
100 plates by S. Moricand, engraved by Heyland or lithographed by Schmidt.
For dates of publication see M. J. van Steenis-Kruseman & W. T. Stearn in *Flora Malesiana* I, 4: ccii (1954).

MORIS, GUISEPPE GIACINTO
FLORA SARDOA SEU HISTORIA PLANTARUM IN SARDINIA ET ADJACENTIBUS INSULIS . . .
3 vols and Atlas. 4to (30 cm × 22 cm). Taurini (Turin) 1837–59.
114 plates (pls 1–111, 33 bis, 77 bis, 78 bis) by M. Lisa and Heyland, engraved by S. Botta, L. Fea, H. Mil and A. Nizza.

MORRIS, RICHARD
FLORA CONSPICUA; A SELECTION OF THE MOST ORNAMENTAL . . . TREES, SHRUBS AND HERBACEOUS PLANTS . . .
8vo (23 cm × 14½ cm). London (1825–) 1826.
60 hand-coloured engraved plates by W. Clark.

MOUILLEFERT, PIERRE
TRAITÉ DES ARBRES ET ARBRISSEAUX FORESTIERS . . . PLUS PARTICULIÈREMENT EN FRANCE . . .

Pl. 83.
Campanula peregrina.
Campanule voyageuse.
Peint par G.W. Voelcker
Gravé par F.W. Meyer

1 vol. text in 2 parts & Atlas. 8vo (23 cm × 15 cm). Paris [1891–] 1892–98.
195 plates (9A–K uncoloured engravings; photographic plates pls 1–144; 40 partly colour-printed, partly hand-coloured, unsigned plates pls 1–32, 26 bis, 26 ter, 27 bis, 28 bis, 28 ter, 29 bis, 32 bis).
For dates of publication, see E. M. Tucker in *J. Arnold Arb.* 3: 227–29 (1922).

MÜLLER, FERDINAND JAKOB HEINRICH VON
THE PLANTS INDIGENOUS TO THE COLONY OF VICTORIA . . . THALAMI-FLORAE . . . LITHOGRAMS.
2 vols & Atlas. 4to (30½ cm × 24 cm). Melbourne 1860–65 (Vol. 1, Atlas); 1910 (Vol. 2 by A. J. Ewart).
122 lithographed plates (pls 1–71, with two each numbered 5 and 10.
Supplemental plates 1–18; pls 1–31 [Vol. 2, 1910]) by F. Schönfeld, L. Becker and T. S. Ralph.

MÜLLER, FERDINAND JAKOB HEINRICH VON
EUCALYPTOGRAPHIA. A DESCRIPTIVE ATLAS OF THE EUCALYPTS OF AUSTRALIA AND THE ADJOINING ISLANDS.
10 parts 4to (30 cm × 24 cm). Melbourne & London 1879–84.
110 plates (11 unnumbered plates in each 'decade') by R. Austen, Rummel and Todt, lithographed by R. Austen.

MÜLLER, FERDINAND JAKOB HEINRICH VON
DESCRIPTION AND ILLUSTRATIONS OF THE MYOPORINOUS PLANTS OF AUSTRALIA . . .
Vol. II. Lithograms 4to (31 cm × 24 cm). Melbourne 1886.
75 plates (frontispiece, pls. 1–72 and Supplemental plates 1, 2) by R. Graff, lithographed by C. Troedel & Co.
Vol. I of text was never published; a text was however, suppled by F. Kränzlin, 'Beiträge zur Kenntnis der Familie Myoporinae R. Br.,' *Fedde Repert, Beih.* 54 (1929).

MÜLLER, FERDINAND JAKOB HEINRICH VON
ICONOGRAPHY OF AUSTRALIAN SPECIES OF ACACIA AND COGNATE GENERA.
13 parts 4to (30 cm × 24 cm). Melbourne. 1887–88.
130 unnumbered plates by R. Graff, lithographed by C. Troedel & Co.

MÜLLER, FERDINAND JAKOB HEINRICH VON
ICONOGRAPHY OF AUSTRALIAN SALSOLACEOUS PLANTS.
9 parts 4to (30 cm × 23½ cm). Melbourne 1889–91.
90 lithographed plates by R. Graff.

MÜLLER, FERDINAND JAKOB HEINRICH VON
ICONOGRAPHY OF CANDOLLEACEOUS PLANTS.
4to (30 cm × 23½ cm). Melbourne 1892.
10 lithographed plates by R. Graff.

MUNTING, ABRAHAM
PHYTOGRAPHIA CURIOSA EXHIBENS ARBORUM, FRUTICUM HERBARUM AC FLORUM ICONES . . .
2 vols Folio (42 cm × 27 cm). Amsterdam and Leyden 1702.
245 uncoloured engravings and frontispiece by J. Goerce, engraved by J. Baptist. H.B.

NEES VON ESENBECK, THEODOR FRIEDRICH LUDWIG
PLANTAE OFFICINALES ODER SAMMLUNG OFFICINELLER PFLANZEN . . .
2 vols. Folio (48 cm × 30½ cm). Düsseldorf (1821–) 1828.
432 hand-coloured lithographed plates by A. Henry, lithographed by Arnz and Co.
Variations are found in the title pages and title is sometimes found as Plantae medicinales . . .
Supplement: ICONES PLANTARUM MEDICINALIUM ODER SAMMLUNG OFFICINELLER PFLANZEN . . . (PLANTAE MEDICINALES . . .).
Folio (48 cm × 30½ cm). Düsseldorf 1833.
120 hand-coloured lithographed plates.

NEES VON ESENBECK, THEODOR FRIEDRICH LUDWIG, & SINNING, WILHELM
SAMMLUNG SCHÖNBLÜHENDER GEWÄCHSE . . . DES KGL. BOTANISCHEN GARTENS ZU BONN . . .
10 parts Folio (Text, 25 cm × 21½ cm); Atlas (46 cm × 29½ cm). Düsseldorf [1825–] 1831.
100 hand-coloured plates by A. Henry, Hohe and Wild, lithographed by Arnz and Co. Plates 1–55 issued unnumbered but numbers are allotted to them in the Register.
Published in 10 parts: pls 1–10 in 1825, pls 11–20 in 1826, pls 21–30 in 1827, pls 31–50 in 1828, pls 51–70 in 1829, pls 71–80 in 1830, pls 81–90 in 1830 or 1831, pls 91–100 in 1831.

PLATE 40
Johann Hoffmannsegg described Portuguese wildflowers in the 110 prints of Flore portugaise *(Berlin 1809–40), made from watercolors by G. W. Voelcker. "Campanula peregrina" exemplifies the almost photographic realism achieved in these handmade stipple etchings. Note in particular the leaves that seem to grow away from the viewer.*

NIETNER, T. & ENDELL, MARIA

DIE ROSE: IHRE GESCHICHTE, ARTEN, KULTUR UND VERWENDUNG . . .
4to (30 cm × 21 cm). Berlin 1880.
12 chromolithographs. H.B.

NUTTALL, THOMAS

THE NORTH AMERICAN SYLVA; OR, A DESCRIPTION OF THE FOREST TREES . . . NOT DESCRIBED IN THE WORK OF F. ANDREW MICHAUX, . . . VOL. I BEING THE FOURTH VOLUME OF MICHAUX AND NUTTALL'S NORTH AMERICAN SYLVA.
8vo (28 cm × 17 cm). Philadelphia 1842 [-43].
40 hand-coloured plates by J. B. Butler, J. T. Worley, lithographed by T. Sinclair.
Later ed. 3 vols 8vo 1852 (being the 4th (-6th) volume of Michaux and Nuttall's North American Sylva).
121 hand-coloured lithographed plates.
Other editions 1854, 1859 (3 vols in 2), 1865.

OEDER, GEORG CHRISTIAN

FLORA DANICA. ICONES PLANTARUM SPONTE NASCENTIUM IN REGNIS DANIAE ET NORVEGIAE, IN DUCATIBUS SLESVICI ET HOLSATIAE, ET IN COMITATIBUS OLDENBURGI ET DELMENHORSTIAE, AD ILLUSTRANDUM OPUS DE IISDEM PLANTIS, REGIO JUSSU EXARANDUM, FLORAE DANICAE NOMINE INSCRIPTUM.
17 vols (51 Heft). Folio (38 cm × 24 cm). Hauniae (Copenhagen) 1761–1883.
3060 hand-coloured engraved plates.
Vol 1–3 and 4. 1 (Heft 1–10) pls 1–600, edited by C. G. Oeder.
Vols 4–5 (Heft 11–18) pls 601–900, edited by F. F. Müller.
Vols 6–7 (Heft 16–21) pls 901–1260, edited by M. Vahl.
Vols 8–13 (Heft 22–39) pls 1261–2340, edited by J. W. Hornemann.
Vols 14 (Heft 40–42) pls 2341–2520, edited by F. M. Liebmann
Vols 15 (Heft 43–45) pls 2521–2700, edited by F. M. Liebmann & J. Lange
Vols 16–17 (Heft 46–51) pls. 2701–3060, edited by J. Lange.
3060 hand-coloured plates by M. Rössler, C. F. Müller, J. T. Bayer, C. Thornam, C. M. Gottsche, T. Jensen, J. F. Kaltenhafer, S. O. Lindberg, Mincke, Picard, Salvetti, Schumacher, engraved by M. Kössler, C. F. Müller, J. T. Bayer, Petersen, C. Thornam, Kaltenhafer, J. Hansen, A. Thornam.
Supplementum Icones plantarum sponte nascentium in Regnis Sueciae et Norvegiae.
Vol. 1 (Fasc. 1–3). 1853–74.
180 hand-coloured plates by J. Bayer, C. Thornam, Gottsche, J. O. Lindberg, engraved by C. Thornam, A. Thornam, U. Hansen. Text of fasc. 1 by F. M. Liebmann, of fasc. 2 and 3 by J. Lange. See J. Anker 'The early history of the Flora Danica', *Libri* (Copenhagen) 1: 334–350 (1951).

OTTO, CHRISTOPH FRIEDRICH. See PFEIFFER, L.G.K. 1838 & LINK, H. F. 1820, 1828, 1840.

PALISOT DE BEAUVOIS, AMBROISE MARIE FRANÇOIS JOSEPH

FLORE D'OWARE ET DE BENIN, EN AFRIQUE.
2 vols (45 cm × 29½ cm). Paris An. XII. 1804 [1805–21].
120 engraved plates by J. G. Prêtre, S. de Luigné and B. Mirbel, engraved by Bouquet, Canu and Lambert. For dates of publication see J. H. Barnhart in *Proc. Amer. Phil. Soc.* 76: 914–920 (1936). H. S. Marshall in *Kew* Bull. 1951, 43–49, M. J. van Steenis-Kruseman & W. T. Stearn in *Flora Malesiana* I, 4: ccv (1954).

PALLAS, PETER SIMON

FLORA ROSSICA SEU STIRPIUM IMPERII ROSSICI PER EUROPAM ET ASIAM INDIGENARUM DESCRIPTIONES ET ICONES . . .
2 vols Folio (49 cm × 30 cm). Petropoli (St. Petersburg), Lipsiae (Leipzig) 1784–88.
101 hand-coloured engraved plates (pls 1–50, 8b, 51–100) by K. F. Knappe. Plates 101–125 were issued in a limited edition without text in 1831; see B. D. Jackson in *J. Bot. (London)* 38: 189 (1900), W. T. Stearn in *J. Arnold Arb.* 22: 229 (1941).

PALLAS, PETER SIMON

SPECIES ASTRAGALORUM DESCRIPTAE ET ICONIBUS COLORATIS ILLUSTRATAE.
Folio (43 cm × 26½ cm). Lipsiae (Leipzig) 1800 [-02].
98 hand-coloured unsigned engraved plates (pls 1–91, 20b, c, d, 43b, 58b, 60b, 70b) by C.G.H. Geissler; see preface, p. viii.

PALLAS, PETER SIMON

ILLUSTRATIONES PLANTARUM IMPERFECTE VEL NONDUM COGNITARUM.
Folio (43½ cm × 28½ cm). Lipsiae (Leipzig). 1803 (-06).
59 unsigned hand-coloured engraved plates by C.G.H. Geissler.

PAUL, WILLIAM

THE ROSE GARDEN.
Royal 8vo. London 1848.
15 hand-coloured lithographs. Later editions up till 1890. H.B.

PAXTON, SIR JOSEPH. See LINDLEY, J. 1882 and Periodicals 1843.

PERROTTET, G. S. See GUILLEMIN, J.B.A. 1830

PEYRITSCH, JOHANN

AROIDEAE MAXIMILIANAE. DIE AUF DER REISE S.M. DES KAISERS MAXIMILIAN I. NACH BRASILIEN GESAMMELTEN ARONGEWÄCHSE. NACH HANDSCHRIFTLICHEN AUFZEICHNUNGEN VON H. SCHOTT BESCHRIEBEN. (S.M. DES KAISERS VON MEXICO MAXIMILIAN I. REISE NACH BRASILIEN (1859–60). BOTANISCHE ERGEBNISSE).
Folio (60 cm × 44 cm). Wien (Vienna) 1879.
Frontispiece and 42 chromolithographed plates by W. Liepolt and J. Selleny, lithographed by A. Hartinger, G. Seelos & M. Streicher.

PEYRITSCH, JOHANN. See KOTSCHY, C.G.T. 1867

PFEIFFER, AUGUST

MAGASIN FÖR BLOMSTER-ÄLSKARE OCH IDKARE AF TRÄGAROS SKÖTSEL.
Folio (29 cm × 22 cm). Stockholm 1803.
36 hand-coloured engravings. H.B.

PFEIFFER, LUDWIG GEORG KARL & OTTO, CHRISTOPH FRIEDRICH.

ABBILDUNG UND BESCHREIBUNG BLÜHENDER CAKTEEN. FIGURES DES CACTÉES EN FLEUR . . . AVEC UN TEXTE EXPLICATIF.
2 vols 4to (33 cm × 25 cm). Cassel [1838–] 1843–50.
60 hand-coloured plates (pls 1–30, 1–30) by J. Prestele, lithographed by B. Dondorf, G. Francke and T. Fischer.
For dates of publication and modern names of plants figured, see W. T. Stearn in *Cactus J.* 8: 39–46 (1939).

PICCIOLI, ANTONIO

L'ANTOTROFIA; OSSIA, LA COLTIVAZIONE DE FIORI
2 vols 8vo (23 cm × 15 cm). Florence 1834.
72 hand-coloured lithographs. H.B.

PIERRE, JEAN BAPTISTE LOUIS

FLORE FORESTIÈRE DE LA COCHIN-CHINE.
5 vols Folio (55 cm × 39 cm). Paris (1879–1907).
400 plates by L. Delpy, lithographed by L. Hugon, J. Storck and E. Oberlin.
Published in 26 parts, for dates of which see the printed list inserted either in Vol. 1 or Vol. 5.

PLANSON, J. A.

ICONOGRAPHIE DU GENRE OEILLET OU CHOIX DES OEILLETS LES PLUS BEAUX ET LES PLUS RARES.
Small Folio. Paris 1845.
50 hand-coloured lithographs showing 200 varieties of Carnations. H.B.

PLÉE, FRANÇOIS

TYPES DE CHAQUE FAMILLE ET DES PRINCIPAUX GENRES DES PLANTES CROISSANT SPONTANÉMENT EN FRANCE . . .
2 vols 4to (27 cm × 20 cm). Paris 1844–64.
160 hand-coloured engraved plates by Plée.

PLENCK, JOSEPH JAKOB

ICONES PLANTARUM MEDICINALIUM SECUNDUM SYSTEMA LINNAEI DIGESTARUM . . . ABBILDUNGEN DER MEDICINALPFLANZEN . . .
8 vols Folio (46 cm × 32½ cm). Viennae 1788–1812.
758 unsigned hand-coloured engraved plates (and title-page to Vol. I engraved by J. I. Albrecht). Of Vol. VIII only 2 parts were issued owing to the death of Plenck in 1807.
Delightful and decorative plates. W.B.

PLUMIER, CHARLES

DESCRIPTION DES PLANTES DE L'AMÉRIQUE AVEC LEUR FIGURES.
Folio (40½ cm × 26½ cm). Paris 1693.
108 engraved plates by C. Plumier.
Reprinted 1713.
Clear, crude figures. W.B.

PLUMIER, CHARLES

PLANTARUM AMERICANARUM . . . QUAS OLIM CAR. PLUMIERIUS . . . DEPINXIT . . . EDIDIT . . . JO BURMANNUS.
10 parts Folio (40 cm × 25½ cm). Amstelaedami (Amsterdam), Lugd. Batav. (Leyden) 1755–60.
263 unsigned engraved plates (pls 1–262, 25*) by C. Plumier.
The originals are at Paris but coloured copies are at the British Museum (Natural History) and University Library, Groningen, the latter being those consulted by Linnaeus.

POEPPIG, EDUARD FRIEDRICH & ENDLICHER, STEPHAN

NOVA GENERA AC SPECIES PLANTARUM QUAS IN REGNO CHILENSI PERUVIANO ET IN TERRA AMAZONICA . . . LEGIT ET DESCRIPSIT ICONIBUSQUE ILLUSTRAVIT.
3 vols Folio (40 cm × 29 cm). Lipsiae (Leipzig) 1835–45.
300 plates by E. Poeppig and J. Zehner, engraved by M. Bauer, A. Bogner, Gebhart and G. Langer.
For dates of publication see W. T. Stearn in *Flora Malesiana* I, 4: ccv (1954).

POHL, JOHANN BAPTIST EMANUEL

PLANTARUM BRASILIAE ICONES ET DESCRIPTIONES HACTENUS INEDITAE.
2 vols Folio (44 cm × 30 cm; 54 cm × 36 cm).

Rosa Indica Cruenta.

Rosier du Bengale à fleurs pourpre de sang.

P. J. Redouté pinx. · Imprimerie de Rémond · Langlois sculp.

Vindobonae (Vienna) [1826–]. 1827–31 [–33].
200 hand-coloured lithographed plates by Sandler.
For dates of publication see W. T. Stearn in *Flora Malesiana* I, 4: ccvi (1954).

POIRET, JEAN LOUIS MARIE. See CHAUMETON, F. P. 1815.

POIRET, JEAN LOUIS MARIE & TURPIN, P.J.F.
LEÇONS DE FLORE. COURS COMPLÈT DE BOTANIQUE, EXPLICATION DE TOUS LES SYSTÈMES.
3 vols 4to (32 cm × 23 cm). Paris 1819–1820.
65 plates printed in colour and finished by hand numbered 1–56. Also 8 additional plates and one large folding plate. Text by Poiret, plates by Turpin. Published in 17 parts. Wrongly advertised in text as having 66 plates. H.B.

POITEAU, PIERRE ANTOINE
FLORA PARISIENSIS . . . PLANTARUM CIRCA LUTETIAM SPONTE NASCENTIUM DESCRIPTIONES, ICONES . . . FLORE PARISIENSE CONTENANT LE DESCRIPTION DES PLANTES ETC.
8 parts Folio (48½ cm × 34½ cm). Paris 1808 [–13].
48 colour-printed partly hand-coloured plates (numbered 1–75, but pls 6, 11, 16, 21, 27, 33, 36, 42–44, 54–58, 60, 62–67, 69–73 not published) by Poiteau and P. J. Turpin, engraved by L. J. Allais, Choubard, Dien, Duvivier, Jacquinot and Melini.
Issued in 8 parts each with 6 plates (fasc. 1–5 in 1808, fasc. 6–7 in 1809) but never completed.

POITEAU, PIERRE ANTOINE. See RISSO, J. A. 1818.

PRATT, ANNE
THE FLOWERING PLANTS . . . OF GREAT BRITAIN AND THEIR ALLIES . . .
5 vols 8vo (23 cm × 14 cm). London [1850–57].
241 colour-printed plates by A. Pratt.
As additional volumes uniform in style were published later *The British Grasses and Ferns*, 35 coloured plates [1859] and *The Ferns of Great Britain and Their Allies* [1855]. This very popular work was many times reprinted.
The most widely-used book on English wild flowers for half a century. A very popular Victorian work. W.B.

PRESL, KAREL BORIWOG
RELIQUIAE HAENKEANAE, SEU DESCRIPTIONES ET ICONES PLANTARUM QUAS IN AMERICA . . . ET IN INSULIS PHILIPPINIS ET MARIANIS COLLEGIT THADDEUS HAENKE.
2 vols Folio (37 cm × 24 cm). Pragae (Prague) 1830 [–35].
72 plates by F. Both, F. Fieber and Longer, engraved by F. Both, J. Skala & W. Zelisko.
For dates of publication, see W. T. Stearn in *J. Soc. Bibl. Nat. Hist.* 1: 153 (1938).

PRÉVOST, JEAN-LOUIS
COLLECTION DES FLEURS ET DES FRUITS PEINTS D'APRÈS NATURE . . . AVEC UNE EXPLICATION DES PLANCHES PAR ANT. NIC. DUCHESNE.
Folio (51 cm × 33½ cm). Paris An. XIII 1805.
48 colour-printed plates by J. L. Prévost, engraved by Chaponnier and L. C. Ruotte.
Discussed in text by Wilfrid Blunt.

PURSH, FRIEDRICH TRAUGOTT
FLORA AMERICAE SEPTENTRIONALIS . . .
2 vols 8vo (22 cm × 13½ cm). London 1814.
24 engraved plates by W. Hooker, in most copies uncoloured, but hand-coloured in a few.
Ed. 2. London 1816.
For date of publication, see W. T. Stearn in *Rhodora* 45: 415, 511 (1943).
An advance copy was presented to Linnean Society of London on 7 Dec. 1813; cf. Graustein in *Rhodora* 56: 275 (1954).

RAOUL, EDOUARD F. A.
CHOIX DE PLANTES DE LA NOUVELLE-ZÉLANDE . . .
Folio (36 cm × 26 cm). Paris 1846.
30 plates by A. Riocreux, engraved by E. Taillant.

RAVENSCROFT, EDWARD JAMES
THE PINETUM BRITANNICUM. A DESCRIPTIVE ACCOUNT OF HARDY CONIFEROUS TREES CULTIVATED IN GREAT BRITAIN.
3 vols Folio (55 cm × 44½ cm). London and Edinburgh [1863–] 1884.
53 hand-coloured lithographed plates by J. Black, W. Richardson, R. K. Greville and J. Wallace, lithographed by R. Black, Day & Son, W. H. MacFarlane, Fr. Schenk; 4 photographs; 643 wood-engraving text-figs. by A. Murray, J. M'Nab, Greville, M. T. Masters. 1 plate containing 3 maps.

PLATE 41
Pierre Joseph Redouté's Les roses *(Paris 1817–24)*
is the jewel of this artist's career. Of its images, "Rosa Indica Cruenta" has a peculiar grace;
comparison with the watercolor shows how perfectly the printmaker has recreated Redouté's original.
In this era of French botanical illustration,
it was the rule to advertise printmakers by engraving their names on the plate.

REDOUTÉ, PIERRE JOSEPH

LES LILIACÉES.

8 vols Folio (53 cm × 35 cm). Paris 1802–1816.

503 plates (numbered 1–486), printed in colour, sometimes finished by hand (see below).

The text of volumes I–IV is by A. P. de Candolle; of volumes V–VII by F. de Laroche and of volume VIII by A. Raffeneau-Delille.

The dates of publication, according to B. B. Woodward in *J. Bot. (London)* 43: 26 (1905) are as follows:

Vol. 1, parts 1–3, pls 1–18, in 1802; part 4, pls 9–24, in 1802 or 1803; parts 5–8, pls 25–48, in 1803; parts 9–10, pls 49–60, in 1804.

Vol. 2, parts 11–17, pls 61–102, in 1804; parts 18–20, pls 103–120, in 1805.

Vol. 3, parts 21–23, pls 121–138, in 1805; parts 24–27, pls 139–162, in 1806; parts 28–30, pls 163–180, in 1807.

Vol. 4, parts 31–34, pls 181–204, in 1807; parts 35–40, pls 205–240, in 1808.

Vol. 5, parts 41–46, pls 241–276, in 1809; parts 47–50, pls 277–300, in 1810.

Vol. 6, parts 51–58, pls 301–348, in 1811; parts 59–60, pls 349–360, in 1812.

Vol. 7, parts 61–67, pls 361–402, in 1812; parts 68–70, pls 403–420, in 1813.

Vol. 8, parts 71–74, pls 421–444, in 1814; parts 75–78, pls 445–468, in 1815; parts 79–80, plates 469–486, in 1816.

The ordinary folio edition, which measures roughly 20 in. by 13 in., began publication in parts in 1802, as noted above. In 1807 publication began of large paper issue (Edition grand papier), measuring roughly 24 in. by 18 in., agreeing with the ordinary folio in text but intended to have the plates retouched by Redouté himself. From vol. 3 onwards the two were issued concurrently. In the preface to the first volume of the large paper issue Redouté wrote: 'J'ai donc attendu les derniers résultats de mes recherches et de mes études sur les moyens d'imiter avec fidélité, par les gravures, les fleurs les plus magnifiques de règne végétal, pour les réunir dans une édition particulière des Liliacées, au nombre de *quarante* exemplaires seulement [*but* dix-huit seulement *according to a pencil note signed* Redouté *in the Lindley Library copy*] tout ce qu'une longue expérience a pu me faire acquérir d'habileté en ce genre. Cette édition différera de celle que le public connait . . . en ce que la première gravure aura participé à tous les perfectionnements . . . elle différera encore en ce que les exemplaires étant très peu nombreux, chaque Liliacée sera retouchée par moi au pinceau, après l'impression'. It will be noted that according to this pencil correction by Redouté there were only 18 and not 40 copies issued of the 'Edition grand papier'. The plates seem to have been issued as they became available without regard to the affinities, classification and nomenclature of the plants. On the completion of the work, however, the plates and text for the large paper issue were re-arranged in alphabetical order by genera. The plate numbers of the ordinary issue, being close to the margin, have sometimes been unintentionally cut away by the binder and those of the large paper issue are sometimes so faint as to be scarcely visible and the leaf of text has no number on the recto.—W. T. STEARN.

REDOUTÉ, PIERRE JOSEPH

LES ROSES. TEXTE PAR CLAUDE ANTOINE THORY.

3 vols Folio (54 cm × 35½ cm). Paris, Imprimerie de Firmin Didot, 1817–1824.

With a portrait of Redouté, engraved by C. S. Pradier after Gérard and 170 plates (including frontispiece), printed in colour and finished by hand.

Some copies contain 168 plates only (plus the frontispiece).

According to B. B. Woodward in *J. Bot. (London)* 43:28 (1905), publication was as follows:

Vol. 1, parts 1–4, pp. 5–72, in 1817; parts 5–9, pp. 73–134, in 1818; part 10–11, pp. 135–158, in 1819.

Vol. 2, parts 12–14, pp. 5–40, in 1819; parts 15–18, pp. 41–92, in 1820; parts 19–21, pp. 93–124, title, in 1821.

Vol. 3, parts 22–23, pp. 5–40, in 1821; parts 24–26, pp. 41–76, in 1822; parts 27–29, pp. 77–108, in 1823; part 30, pp. 109–128, 1–4, in 1824.

Folio edition: on vélin paper, with uncoloured plates on a paper identical to the pages of the printed text and a set of the plates, printed in colour on white vélin, the text being sometimes printed on coloured paper.

Small edition of a few copies only, printed for Treuttel et Wurtz, some copies with the text in English.

First 8vo edition, 3 vols Paris, Panckoucke, 1824–1826.

With 160 plates, printed in colour, a few copies on vélin.

The frontispiece of the larger edition does not figure here.

Second 8vo edition: Décrites et classées selon leur ordre naturel, par Cl. Ant. Thory, 3e édition, publiée sous la direction de M. Pirolle.

3 vols 8vo Paris, de l'imprimerie Crapelet, Dufart 1828–30.

With 2 portraits and 181 plates, printed in colour and finished by hand.

Most complete edition which contains a new methodical classification of roses.

Another edition: Paris 1835. Plates as above.

REDOUTÉ, PIERRE JOSEPH

ALBUM DE REDOUTÉ.

Folio. Paris 1824.

With from 24–30 plates, printed in colour from *Les Roses* and *Les Liliacées, Plantes Grasses* and *Jardin de la Malmaison*.

Complete copies of this book are found with varying numbers of plates from 24–30, nor are the plates themselves always the same.

REDOUTÉ, PIERRE JOSEPH
CHOIX DE QUARANTE PLUS BELLES FLEURS, TIRÉES, DU GRAND OUVRAGE DES LILIACÉES.
Folio. Paris 1824.
40 coloured plates.

REDOUTÉ, PIERRE JOSEPH
CHOIX DES PLUS BELLES FLEURS PRISES DANS DIFFÉRENTES FAMILLES DU RÈGNE VÉGÉTAL ET DE QUELQUES BRANCHES DES PLUS BEAUX FRUITS, GROUPÉES QUELQUEFOIS, ET SOUVENT ANIMÉES PAR DES INSECTES ET DES PAPILLONS.
Large 4to and folio (34 cm × 24½ cm) the latter a 'grand papier' edition. Paris 1827 [–1833].
Published in 36 fascicules, with 144 plates, printed in colour and finished by hand; for dates of publication, see B. B. Woodward in *J. Bot. (London)* 43:29 (1905).
Another edition: Paris 1829.
New edition: Choix des plus belles fleurs et des plus beaux fruits.
Paris n.d. (1833).
Under this title 12 plates by Redouté were published at Paris in 1939 in portfolio. Avant-propos by Colette.
This later edition is much inferior to the first.

REDOUTÉ, PIERRE JOSEPH
CATALOGUE DE 486 LILIACÉES ET DE 168 ROSES, PEINTES PAR P. J. REDOUTÉ.
8vo. Paris 1829.
No illustrations but a list of all the plates in the 'Roses' and the 'Liliacées' with the names in Latin and French.

REDOUTÉ, PIERRE JOSEPH
RECUEIL DE SIX BEAUX BOUQUETS. LITHOGRAPHIES PAR POINTEL DU PORTAIL D'APRÈS LES DESSINS ORIGINAUX DE P. J. REDOUTÉ.
4to. Paris 1835.
6 uncoloured lithographs. Lithography by Villain.

REDOUTÉ, PIERRE JOSEPH
REUNION DE DOUZE PLANCHES DE FLEURS LITHOGRAPHIÉES PAR A. PRÉVOST ET POINTET DU PORTAIL.
4to. Paris n.d. (1835).
12 uncoloured lithographs. Lithography by Villain.

REDOUTÉ, PIERRE JOSEPH
COLLECTION DE JOLIES PETITES FLEURS CHOISIES PARMI LES PLUS GRACIEUSES PRODUCTIONS DE CE GENRE, TANT EN EUROPE QUE DANS LES AUTRES PARTIES DU MONDE.
4to. Paris, Emile Lecomte 1835.
With 48 plates, printed in colour and finished by hand.
Another edition:
4to. Paris, Pillet 1836.
With 28 plates, printed in colour.

REDOUTÉ, PIERRE JOSEPH
CHOIX DE SOIXANTE ROSES.
Folio. Paris, chez l'auteur 1836.
With 60 plates, printed in colour, of new roses, not yet described and an introduction by Jules Janin.
Another imprint: Paris, chez Danlos.

REDOUTÉ, PIERRE JOSEPH
LE BOUQUET ROYAL. OEUVRE POSTHUME DE P. J. REDOUTÉ, DÉDIÉ A S.M. LA REINE DES FRANÇAIS.
Folio. Paris (1843).
4 plates (all of roses) printed in colour and finished by hand.
Other Paris imprints, 1843 and 1844.

REDOUTÉ, PIERRE JOSEPH & PRÉVOST, A.
BOUQUETS.
Folio. Paris 1843.
With 4 uncoloured plates.

REDOUTÉ, PIERRE JOSEPH
FLEURS. ALBUM.
8vo. 1843.

REDOUTÉ, PIERRE JOSEPH
ROSES.
Folio.
31 plates after Redouté, engraved by Charlin, printed in colour, published after the 1828–1830 edition.

REDOUTÉ, PIERRE JOSEPH
LES MOIS.
Folio. Paris n.d.
12 lithographs, coloured by hand.
Bouquets of flowers, one for each month. Very rare.

REDOUTÉ, PIERRE JOSEPH
LA COURONNE DES ROSES
Folio. Paris n.d.
Choix de trente roses, tirées de la monographie de M. Redouté. Coloriées sous les yeux de l'auteur.
30 of the best plates from the *Roses* (q.v.) coloured under the supervision of Redouté himself.

REDOUTÉ, PIERRE JOSEPH
PETITS MODÈLES DE FLEURS.
Oblong 4to. Paris, n.d.
6 uncoloured lithographs. Lithography by Derebergue.

REDOUTÉ, PIERRE JOSEPH
LE COURS DE FLEURS DU JARDIN DES PLANTES, PAR M. P. J. REDOUTÉ ET MMES A. JANET, DE CHANTEREINE, A. WASSET, R. BESSIN, O. ARSON [AND OTHERS].
Folio. Paris n.d.
8 parts of 6 lithographs each, uncoloured, of which 4 by Redouté. Another imprint: Chavan n.d.

Rosa Gallica Versicolor. *Rosier de France à fleurs panachées.*

P. J. Redouté pinx. *Imprimerie de Rémond* *Langlois sculp.*

REDOUTÉ, PIERRE JOSEPH

COLLECTION DE BEAUX BOUQUETS LITHOGRAPHIES PAR DIVERS ARTISTES D'APRÈS LES MEILLEURS PEINTRES DE FLEURS.

Folio. Paris: François de la Rue, 18 Rue J. J. Rousseau. London: Gambart and Co. n.d. (1845).

Title-page and 24 hand-coloured lithographs, 22 after Redouté, by various lithographers, including Grobon Frères and A. Prévost.

There is no other bibliographical record of this book. The plates are *not* copies of those that appear in any other Redouté work. The copy seen is in the library of Major the Hon. Henry Broughton. H.G.

REGNAULT, NICOLAS FRANÇOIS

LA BOTANIQUE MISE À LA PORTÉE DE TOUT LE MONDE . . .

3 vols Folio (47 cm × 34 cm). Paris [1770–] 1774 [–80].

Title-page and 472 hand-coloured engraved plates by F. Regnault and C. de Wangis-Regnault.

A very impressive work. W.B.

REICHENBACH, HEINRICH GOTTLIEB LUDWIG

MONOGRAPHIA GENERIS ACONITI, ICONIBUS OMNIUM SPECIERUM COLORATIS ILLUSTRATA, LATINE ET GERMANICE ELABORATA.

1 vol Folio (43½ cm × 26 cm. Text and engraved plates; 43½ cm × 29 cm, lithographed plates). Lipsiae (Leipzig) 1820–21.

19 hand-coloured plates (Tab A uncoloured, pls 1–18, of which 7 are engraved and 11 lithographed) by F. Guimpel and H.G.L. Reichenbach, engraved by J. F. Schröter and lithographed by C. Meinhold.

According to W. T. Stearn in *J. R. Hort. Soc.* 67:297 (1942), published as follows: pp. i–iv, 1–72, pls 1–6 in 1820, pp. 73–100, pls 7–18 in 1821.

REICHENBACH, HEINRICH GOTTLIEB LUDWIG

MAGAZIN DER AESTHETISCHEN BOTANIK ODER ABBILDUNG UND BESCHREIBUNG DER FUR GARTENCULTURE EMPFEHLUNG-WERTHEN GEWÄCHSE . . . ICONES ET DESCRIPTIONES PLANTARUM CULTARUM ET COLENDARUM.

2 vols 4to (24 cm × 18 cm). Leipzig [1821–] 1822–26.

96 hand-coloured plates by H.G.L. Reichenbach, engraved by Berger, Erdmann, A. Harzer, Richter, C. Schnorr, F. Täubert (Teubert), Wirani. Vol. 1 comprises pls. 1–72, titlepage and index; it was published in 12 parts, the contents of parts 1–6 (pls 1–36; 1821–22) being listed in *Flora (Regensburg)* 6: 129–137 (March 1823), of parts 9–12 (pls 37–12) in *Flora (Regensburg)* 8.i. Beil: 52–60 (1825).

Vol. 2 never completed, comprises pls 73–96.

Publication of the whole work as follows: parts 1–4, pls 1–24, in 1821; parts 5–6, pls 25–36, in 1822; parts 7–11, pls 37–54, in 1823; parts 12–14, pls 67–84, in 1824; part 15, pls 85–90, in 1825; part 16, pls 91–96, in 1826.

REICHENBACH, HEINRICH GOTTLIEB LUDWIG

ILLUSTRATIO SPECIERUM ACONITI GENERIS, ADDITIS DELPHINIIS QUIBUSDAM. NEUE BEARBEITUNG DER ARTEN DER GATTUNG ACONITUM, UND EINIGER DELPHINIEN.

Folio (35 cm × 23 cm). Lipsiae (Leipzig). 1823–27.

71 hand-coloured engraved and 1 uncoloured lithographed plate by H.G.L. Reichenbach, engraved by A. Harzer, C. Schnorr, T. Täubert and G. Zumpe.

According to W. T. Stearn in *J. R. Hort. Soc.* 67: 297 (1942) pls 1–24 in 1823, pls 25–36 in 1824, pls 37–54 in 1825, pls 55–72 in 1827.

REICHENBACH, HEINRICH GOTTLIEB LUDWIG

ICONOGRAPHIA BOTANICA, SEU PLANTAE CRITICAE. ICONES PLANTARUM RARIORUM . . . KUPFERSAMMLUNG KRITISCHE GEWÄCHSE . . .

9 vols 4to (24 cm × 19 cm). Lipsiae (Leipzig) 1823–32.

1000 hand-coloured or uncoloured plates by H.G.L. Reichenbach, Hümizsch and Kätzing, engraved by A. Harzer, J. and C. Schnorr and others.

REICHENBACH, HEINRICH GOTTLIEB LUDWIG

ICONOGRAPHIA BOTANICA EXOTICA SIVE HORTUS BOTANICUS, IMAGINES PLANTARUM IMPRIMIS EXTRA EUROPAM INVENTARUM COLLIGENS . . . KUPFERSAMMLUNG DER . . . AUSLÄNDISCHEN GEWÄCHSE . . .

3 vols 4to (25 cm × 19½ cm). Lipsiae (Leipzig). 1827–30 [1844].

250 hand-coloured engraved and lithographed plates by H.G.L. Reichenbach and Humm, engraved by F. Guimpel, A. Harzer, Hase, Hilscher, C. Schnorr and Zumpe or lithographed by Htzsch.

PLATE 42

"Rosa Gallica Versicolor" illustrates Pierre Joseph Redouté's Les roses *(Paris 1817–24). Empress Josephine maintained huge gardens of exotic flowers but held a special affection for roses. Redouté owed his career to Josephine's patronage, and he dedicated this work to her. It idealizes the form of the monograph with 168 stunning color-printed stipple etchings and an excellent text by Claude Antoine Thory.*

REICHENBACH, HEINRICH GOTTLIEB LUDWIG

FLORA EXOTICA. DIE PRACHTPFLANZEN DES AUSLANDES IN NATURGETREUEN ABBILDUNGEN.

5 vols Folio (35½ cm × 27 cm). Leipzig. 1834–36.

360 unsigned unnumbered hand-coloured lithographed plates by H.G.L. Reichenbach.

REICHENBACH, HEINRICH GOTTLIEB LUDWIG & OTHERS

ICONES FLORAE GERMANICAE ET HELVETICAE . . . DEUTSCHLANDS FLORA MIT HÖCHST NATURGETREUEN CHARAKTERISTISCHEN ABBILDUNGEN ALLER IHRER PFLANZENARTEN IN NATÜRLICHEN GRÖSSE . . .

Vols 1–11 bear also the title: *Iconographia Botanica seu plantae criticae*, of which they form cents. xi–xx, Vols 13–21 are by H. G. Reichenbach (filius) 25 vols (of which 22–25 were published after 1900 and are larger). 4to (26 cm × 20 cm). Leipzig. 1834–67 (Vols 1–21).

2719 hand-coloured plates by H.G.L. and H. G. Reichenbach, and C. H. Schnorr, engraved by C. H. Schnorr, Berger, A. Harzer, A. Weidenbach, Berthold, Gebhardt, Scherell, A. G. Schwerdgeburth, and Werner.

Vol.		
1	pls	1–110 (1834).
2	pls	1–102 (1837–38).
3–4	pls	1–19, 1–40, 1–128, 16bis, i.e. total 188 (1838–40).
5–6	pls	129–352, 198bis, 277bis, 282bis, i.e. total 227 (1841–44).
7	(suppl.)	1–82, 51b, i.e. total 83 (1845).
8	pls	193–318, i.e. total 126 (1846).
9	pls	319–418, 324b, 417 missing. i.e. total 100 (1847).
10	pls	419–520, i.e. total 102 (1848).
11	pls	521–620, i.e. total 100 (1849).
12	pls	621–731, i.e. total 111 (1850).
13–14	pls	353–522 (top left hand corner, 1–170), i.e. total 170 (1850–51).
15	pls	732–891 (top left hand corner, 1–160) i.e. total 160 (1852–1853).
16	pls	892–1041 (top left hand corner, 1–150) i.e. total 150 (1853–54).
17	pls	1042–1201 (with numbers 1117–1126 missing, the others numbered in top left hand corner 1–150) i.e. total 150 (1854–55).
18	pls	1202–1351 (top left hand corner 1–150) i.e. total 150 (1856–58)
19 I	pls	1352–1621 (with numbers 1441–1450 missing, the others numbered in top left-hand corner 1–260) i.e. total 260 (1858–60).
20	pls	1622–1841 (top left hand corner 1–220) i.e. total 220 (1861–62).
21	pls	1842–2051 (top left hand corner 1–210) i.e. total 210 (1863–67).

In 1898 publication was resumed at Leipzig and Gera, under the editorship of F. C. Kohl and then continued by G. Beck von Mannagetta, with mostly lithographed plates, until the First World War.

Vol.		
19II	pls	1–81 (1904–06).
III	pls	175–308 (1909–12).
22	pls	1–220 (1867–89).
	pls	221–272, 148*, 160*, 164*, 169*, 178*, 189*–191*, 193*, 195*, 220*, 220**, total 284 (1900–03).
23	pls	1–138, 64a, 74a, 119a, 138a, 138b, total 143 (1896–99).
24	pls	139–301, 151* total 164 (1903–09).
25I	pls	1–79 (1909–12).
II	pls	80–119 (1913–14).

A 2nd Ed. of Vol. 1 (*Agrostographia Germanica*) was published in 1850.

REICHENBACH, HEINRICH GUSTAV

XENIA ORCHIDACEA. BEITRÄGE ZUR KENNTNISS DER ORCHIDEEN.

3 vols 4to (27½ cm × 20½ cm). Lipsiae (Leipzig) (1854–) 1858–1900.

300 engraved plates (some partly hand-coloured) by H. G. Reichenbach, Wagener, Kränzlin and others, engraved by Berthold, Scherell, W. Werner and others.

For dates of publication see vol. 1: 240 (1858), 2: 224 (1874), 3: 192 (1900); for vol. 3: 65–192, pls. 231–300 (1890–1900) F. Kränzlin was responsible.

REIDER, JACOB ERNST VON

ANNALEN DER BLUMISTEREI.

12 vols 12mo (20 cm × 12 cm). Nuremberg and Leipzig 1825–1836.

288 hand-coloured engravings. H.B.

REITTER, JOHANN DANIEL & ABEL, GOTTLIEB FRIEDRICH

ABBILDUNG DER HUNDERT DEUTSCHEN WILDEN HOLZ-ARTEN NACH DEM NUMMERN-VERZEICHNIS IM FORST-HANDBUCH VON F. A. L. VON BURGSDORF . . .

4 parts 4to (27½ cm × 22 cm). Stuttgart 1790–94.

100 unsigned hand-coloured engraved plates by G. F. Abel.

REITTER, JOHANN DANIEL & ABEL, GOTTLIEB FRIEDRICH

BESCHREIBUNG UND ABBILDUNG DER IN DEUTSCHLAND SELTENER WILDWACHSENDEN UND EINIGER BEREITS NATURALISIERTEN HOLZARTEN ALS

FORTSETZUNG VON DEN ABBILDUNGEN . . . NACH DEM NUMMERN-VERZEICHNISS IM FORSTHANDBUCH VON F. A. L. VON BURGSDORF . . .
4to (27½ cm × 22 cm). Stuttgart 1803.
25 unsigned hand-coloured engraved plates by G. F. Abel.

RICHARD, ACHILLE

TENTAMEN FLORAE ABYSSINICAE SEU ENUMERATIO PLANTARUM HUCUSQUE IN PLERISQUE ABYSSINIAE PROVINCIIS DETECTARUM ET PRAECIPUE A RICH. QUARTIN DILLON ET ANT. PETIT (ANNIS 1838–43) LECTARUM.
2 vols Text, 8vo (25 cm × 16 cm). Icones (Atlas), Folio (50½ cm × 34 cm). Paris [1847–51].
103 engraved plates (1–102, 53bis) by Vauthier, engraved by Massard, Annedouche and others. This forms part 3, Vols 4 and 5, of the Histoire naturelle, Botanique of VOYAGE EN ABYSSINIE . . . par T. Lefebvre, A. Petit et Q. Dillon [et] Vignaud.

RICHARD, ACHILLE. See GUILLEMIN, J. B. A. & OTHERS. 1831.

RISSO, J. ANTOINE & POITEAU, PIERRE ANTOINE

HISTOIRE NATURELLE DES ORANGERS.
Folio (47 cm × 30½ cm). Paris 1818 [–20].
109 colour-printed plates by Poiteau, engraved by V. Bonnefoi, Chailly, Dien, Gabriel, Legrand, T. Susémihl and Texier.

RISSO, J. ANTOINE & POITEAU, PIERRE ANTOINE

HISTOIRE ET CULTURE DES ORANGERS.
Folio (35 cm × 26½ cm). Paris 1872.
Nouvelle édition . . . augmentée . . . par A. Du Breuil.
110 colour-printed engraved plates (Nos 1–109, 29bis) by Poiteau.

ROBERT, NICOLAS. See DODART, DENIS, 1701.

ROBLEY, AUGUSTA J.

A SELECTION OF MADEIRA FLOWERS . . .
Folio (45 cm × 32 cm). London 1845.
8 hand-coloured lithographs. H.B.

ROEMER, JOHANN JAKOB

FLORA EUROPAEA INCHOATA.
14 parts, 8vo (20 cm × 11½ cm). Norimbergae (Nuremberg) 1797–1811.
112 hand-coloured engraved plates by Louise Roemer, engraved by Vogel.

ROESSIG, CARL GOTTLOB

DIE ROSEN NACH DER NATUR GEZEICHNET UND COLORIRT . . . LES ROSES . . . TRADUIT DE L'ALLEMAND PAR M. DE LAHITTE.
2 vols 4to (31 cm × 25 cm). Leipzig [1802–20].
60 hand-coloured engraved plates by Luise von Wangenheim.

ROSCOE, MRS EDWARD (MARGARET LACE).

FLORAL ILLUSTRATIONS OF THE SEASONS . . .
4to (28½ cm × 22 cm). London 1829–31.
55 aquatint plates by Mrs Roscoe, engraved by R. Havell jun.
This was issued with one of two title-pages as follows:
1st. Floral illustrations of the seasons, consisting of the most beautiful, hardy and rare herbaceous plants, cultivated in the flower garden, from drawings by Mrs Edward Roscoe, Liverpool, engraved by R. Havell jun. London. 1831.
2nd Floral illustrations of the seasons, consisting of representations drawn from nature of some of the most beautiful, hardy and rare herbaceous plants cultivated in the flower garden, carefully arranged according to their seasons of flowering, with botanical descriptions, directions for culture etc: London. 1829.

ROSCOE, WILLIAM

MONANDRIAN PLANTS OF THE ORDER SCITAMINEE, CHIEFLY DRAWN FROM LIVING SPECIMENS . . .
2 vols Folio (53½ cm × 42 cm). Liverpool [1824–] 1828 [–29].
112 unsigned hand-coloured lithographed plates by T. Allport, Mrs J. Dixon, E. Fletcher, R. Miller, Mrs E. Roscoe, M. Waln and E. Yates, under the direction of G. Graves Jun.
For some information about dates of publication, see W. T. Stearn in *Flora Malesiana* I, 4: ccx (1954).

ROSENBERG, MARY ELIZABETH

CORONA AMARYLLIDACEAE.
Large Folio (46 cm × 34 cm). London n.d. (1835).
8 water-colour drawings with printed title page and text.
Two copies of this have been seen so it is evidently a real book like the first edition of Lewin's *Birds*, and Mrs Bowdich's *Fishes*, in both of which the plates were all water-colour drawings. H.B.

ROSENBERG, MARY ELIZABETH

CORONA AMARYLLIDACEAE.
4 parts. Folio (45½ cm × 34½ cm). Bath 1839.
8 hand-coloured plates by M. E. Rosenberg.

ROSENBERG, MARY ELIZABETH

THE MUSEUM OF FLOWERS.
Royal 8vo. London 1845.
54 hand-coloured lithographs. H.B.

ROTH, ALBRECHT WILHELM

CATALECTA BOTANICA, QUIBUS PLANTAE NOVAE ET MINUS COGNITAE DESCRIBUNTUR ATQUE ILLUSTRANTUR.
3 parts, 8vo (20 cm × 11½ cm). Lipsiae (Leipzig) 1797–1806.
29 hand-coloured engraved plates (1–8, 1–9, 1–12) by C. F. Mertens, J. Sturm, Hayne, engraved by J. Sturm, W. Stadelmann, C.J.C. Wirsing.

[ROUPELL, ARABELLA E. (MRS T. B. ROUPELL)].

SPECIMENS OF THE FLORA OF SOUTH AFRICA, BY A LADY.
Folio (57½ cm × 45 cm). [London 1849].
10 hand-coloured plates (including title-page) by A. E. Roupell, under the direction of Wallich and W. J. Hooker, lithographed by P. Gauci.

ROUSSEAU, JEAN JACQUES

LA BOTANIQUE . . .
Folio (55 cm × 36 cm) and 4to. Paris An XIV–1805.
65 colour-printed plates by P. J. Redouté, engraved by Bouquet, J. Chailly, Mlle Delélo, Gabriel, de Gouy, J. Marchand, Masson, Prot, Souet, Suel & Tassaert.
3rd edition. Paris 1821.

ROXBURGH, WILLIAM

PLANTS OF THE COAST OF COROMANDEL; SELECTED FROM DRAWINGS AND DESCRIPTIONS . . .
PUBLISHED . . . UNDER THE DIRECTION OF SIR JOSEPH BANKS.
3 vols Folio (57½ cm × 45 cm). London 1795–1819.
300 hand-coloured plates by various Indian artists, engraved by Mackenzie, Girtin, R. B. Peake and Weddell.
For dates of publication, see W. T. Stearn in *Flora Malesiana* I, 4: CCX (1954).

ROYLE, JOHN FORBES

ILLUSTRATIONS OF THE BOTANY AND OTHER BRANCHES OF THE NATURAL HISTORY OF THE HIMALAYAN MOUNTAINS, AND OF THE FLORA OF CASHMERE.
2 vols Folio (36 cm × 26 cm). London (1833–) 1839 [–40].
Engraved frontispiece by R. Smith, engraved by J. Clark, and 100 mostly hand-coloured plates (including Nos. 75a, '63a or 79', '84a or 98', '99 or 78a', '83 or 100', no plate is numbered '76' but the plate issued as '75' should be 76 and the plate issued as '75a' should be 75) by Vishnupersaud, Capt. Cantley, C. M. Curtis, Miss Drake, J. T. Hart, W. Saunders, Luchmun Sing, J.D.C. Sowerby and J. O. Westwood, lithographed by M. Gauci.
For dates of publication see W. T. Stearn in *J. Arnold Arb.* 24: 484 (1943).

RUDGE, EDWARD

PLANTARUM GUIANAE RARIORUM ICONES ET DESCRIPTIONES HACTENUS INEDITAE.
4 parts Folio (43½ cm × 28½ cm). Londini 1805 (–07).
50 plates by Anne Rudge, engraved by Warner.

RUIZ LOPEZ, HIPŌLITO & PAVON, JOSÉ

FLORA PERUVIANA ET CHILENSIS, SIVE DESCRIPTIONES, ET ICONES PLANTARUM PERUVIANARUM ET CHILENSIUM . . .
3 vols Folio (43½ cm × 31½ cm). [Madrid] 1798–1802.
325 (usually uncoloured but sometimes hand-coloured) engraved plates by J. Brunete, I. Galvez, F. Pulgar, J. Rivera, J. Rubio, engraved by A. Blanco and many others.
Plates for a fourth volume, numbered 326–425, are in a few libraries but do not appear to have been published.

SAGRA, RAMON DE LA

ICONES PLANTARUM IN FLORA CUBANA DESCRIPTARUM EX HISTORIA PHYSICA, POLITICA ET NATURALI A R. DE LA SAGRA EDITA EXCERPTAE.
Folio (42 cm × 27 cm). Parisiis, Londini, New York 1863.
122 uncoloured plates (Crypt: pls 1–20; Plant. Vasc.: pls 1–89, 12bis, 28bis, 36bis, 40bis, 44(1), 44(2), 44(3), 47bis, 49bis, 54bis, 54ter, 59bis), by J. Gontier, Vauthier, and A. Riocreux, engraved by Annedouche and many others.
The Atlas Volume (4) of De la Sagra's *Flora Cubana* (1853), is identical in contents, though slightly smaller in overall size, with *Icones Plantarum Excerptae* (1863). In the former, at Kew, plates I–XX (Crypt) are hand-coloured.
For a list of plates, see I. Urban, *Symb. Antill.* I: 144–5 (1898).

SAINT-HILAIRE, AUGUSTIN FRANÇOIS CÉSAR PROUVENÇAL DE

HISTOIRE DES PLANTES LES PLUS REMARQUABLES DU BRÉSIL ET DU PARAGUAY . . .
1 vol. 4to (29 cm × 22 cm). Paris 1824.
30 engraved plates by Blanchard and Richard (pl. 15 only).
According to O. Kuntze, *Rev. Gen. Pl.* I, CXLIII (1891), this was published in 1825.

SAINT-HILAIRE, AUGUSTIN FRANÇOIS CÉSAR PROUVENÇAL DE

PLANTES USUELLES DES BRASILIENS . . .
14 parts 4to (26 cm × 21 cm). Paris. 1824 [–28].
70 plates by E. Blanchard, lithographed by Langlumé.
For dates of publication, see B. B. Woodward in *J. Bot. (London)* 42: 86 (1904).

SAINT-HILAIRE, AUGUSTIN CÉSAR FRANÇOIS PROUVENÇAL DE

FLORA BRASILIAE MERIDIONALIS.

3 vols 4to (32½ cm × 24 cm). Parisiis (Paris) 1825–33.
192 plates (pls 1–191, two plates with the number 63, 67bis, but 160 missing) by E. Delile and Turpin, engraved by Bourey, Coignet, E. Cornu, David, Dien, Dondey, Georger, Lejeune, A. Massard, C. Noiret, Plée père et fils, Mme. Rebel, E. Taillant.
Two plates bear the number 63: that labelled *Kielmeyera humifusa* accords with the text; that labelled *Larnottea tomentosa* should have been cancelled and replaced by an amended version, 67bis, under the name *Caryocar brasiliense*.

SALISBURY, RICHARD ANTHONY

ICONES STIRPIUM RARIORUM DESCRIPTIONIBUS ILLUSTRATAE.

Folio (57½ cm × 45½ cm). Londini 1791.
10 water-colour drawings by R. A. Salisbury.

SALISBURY, RICHARD ANTHONY

THE PARADISUS LONDINENSIS OR COLOURED FIGURES OF PLANTS CULTIVATED IN THE VICINITY OF THE METROPOLIS. THE DESCRIPTIONS BY RICHARD ANTHONY SALISBURY . . . THE FIGURES BY WILLIAM HOOKER.

2 vols 4to (28 cm × 22 cm). London 1805–1807 [–08].
119 hand-coloured plates (pls. 1–117, and two plates numbered 14 and 67), drawn, engraved and coloured by W. Hooker.
Vol. 1, pls 1–70 (1805–07), Vol. 2, pls 71–117 (1807–08); originally published in monthly parts.
W. Hooker was an able artist, not related to Sir William Jackson Hooker. He did much work for the Horticultural Society of London. W.B.

SALM-REIFFERSCHEIDT-DYCK, JOSEPH FÜRST VON

MONOGRAPHIA GENERUM ALOES ET MESEMBRYANTHEMI.

7 parts 4to (31½ cm × 24 cm). (Fasc. 1–4). Düsseldorf 1836–42.
(Fasc. 5–7). Bonnae (Bonn) 1849–63.
352 unsigned partly hand-coloured plates by Salm-Dyck lithographed by Arnz & Co and A. Henry.
For index to plates with their dates of publication and modern names, see W. T. Stearn in *Cactus J.* 7; 34–44, 66–85 (1938–39).

SANDER, HENRY FREDERICK CONRAD

REICHENBACHIA. ORCHIDS ILLUSTRATED AND DESCRIBED.

1st series 2 vols Folio (52 cm × 39 cm). St. Albans [1886–] 1888–90.
2nd Series 2 vols Folio (52 cm × 39 cm). [1891–] 1892–94.
192 chromolithographed plates (96 in each series) by H. G. Moon, W. H. Fitch and A. H. Loch, lithographed by G. Leutzsch, J. Mansell and J. Macfarlane.
An important and authoritative work for orchid growers.

SARGENT, CHARLES SPRAGUE

THE SILVA OF NORTH AMERICA, A DESCRIPTION OF THE TREES WHICH GROW NATURALLY IN NORTH AMERICA EXCLUSIVE OF MEXICO.

14 vols 4to (36½ cm × 27½ cm). Boston and New York 1891–1902.
740 uncoloured plates by C. E. Faxon, engraved by P. and E. Picart and others under the direction of A. Riocreux.
'Faxon united accuracy with graceful composition and softness of outline'. (C. S. Sargent).

SARGENT, CHARLES SPRAGUE

FOREST FLORA OF JAPAN.

Folio (31 cm × 24 cm). Boston and New York 1894.
26 plates (17 engraved by C. E. Faxon, 2 pen and ink drawings by E. J. Meeker and 7 photographs). Originally published as a series of articles 'Notes on the forest flora of Japan', *Garden and Forest* 6: 26–28, etc. (1893); the 1894 edition in book form has additional illustrations.

SAUNDERS, WILLIAM WILSON. See Periodicals.

SAVI, GAETANO

FLORA ITALIANA, OSSIA RACCOLTA DELLA PIANTE PIU BELLE, CHE SI COLTIVANO NEI GIARDINI D'ITALIA.

3 vols Folio (44 cm × 30½ cm). Pisa 1818–24.
120 engraved plates (stated to be partly colour-printed in some copies) by A. Serantoni.

SAYER, ROBERT & OTHERS

THE FLORIST. CONTAINING SIXTY PLATES OF THE MOST BEAUTIFUL FLOWERS REGULARLY DISPOSED IN THEIR SUCCESSION OF BLOWING . . .

Small 4to (23½ cm × 17 cm) varying considerably in different copies. London *c.* 1760.
60 engraved black and white plates and engraved title-page by C. Phillips and others. Instructions for hand colouring the plates were included and so copies with coloured plates are not infrequently found.
2nd. ed. differing in the list of publishers. n.d.
Reissued as BOWLES'S FLORIST but with the plates re-engraved and several substitutions made, 1774–1777, published by Carington Bowles.

Pl. 16.

AMARYLLIS CROCATA,

Drawn by Mrs. E. Bury Liverpool. PAPILIO NESTOR BRAZIL. Engraved, Printed, & Coloured by R. Havell London.

SCHMIDEL, CASIMIR CHRISTOPH

ICONES PLANTARUM ET ANALYSES PARTIUM AERI INCISAE ATQUE VIVIS COLORIBUS INSIGNITAE . . . CURANTE ET EDENTE GEORG WOLFG. KNORR.

Folio. Norimbergae (Nüremberg) 1747.

50 plates by N. Gabler and J. C. Keller engraved by G. W. Knorr. This first issue has not been examined.

Folio (41 cm × 27 cm). [Nuremberg], 1762–76 (Manip. 1–2); Erlangae [Erlangen] 1787 (Manip. 3). [Manipulus 1] curante et edente J. C. Keller. [i.e. pls 1–25]; Manipulus 2, curante et edente V. Bischoff [i.e. pls 26–50]; Manipulus 3 [i.e. pls 51–75] ed. C. D. Schreber.

75 hand-coloured plates by N. Gabler and J. C. Keller and V. Bischoff, engraved by J. C. Keller and V. Bischoff.

Editio secunda. Folio (42 cm × 26 cm). Erlangae. 1793–97 (Manip. 1–3).

75 hand-coloured engraved plates as in 1762–87 issue.

Some effective plates by Gabler. W.B.

SCHNEEVOOGT, G. VOORHELM

ICONES PLANTARUM RARIORUM DELINEAVIT ET IN AES INCIDIT HENRICUS SCHWEGMAN; . . . SCRIPTIONEM INSPEXIT S. J. VAN GEUNS. AFBEELDINGEN . . . BLOEM-EN PLANT-GEWASSEN . . .

2 vols Folio (45½ cm × 28 cm). Haerlem (Haarlem) [1792–] 1793 [–95].

48 hand-coloured engraved plates all by P. van Loo and H. Schwegman. Text in Latin, Dutch, French and German.

Vol. 1, pls 1–36, published 1792–93; of Vol. 2, only 4 parts (containing pls 37–48) published in 1894–95. For dates of publication see W. T. Stearn in *J. Bot. (London)* 78: 66–74 (1940).

The finest 18th century Dutch botanical work, illustrates plants grown in the famous Voorhelm and Schneevoogt nursery at Haarlem. W.B.

SCHOTT, HEINRICH WILHELM, & ENDLICHER, STEPHAN LADISLAUS.

MELETEMATA BOTANICA.

Folio (47 cm × 31 cm). Vindobonae (Vienna) 1832.

5 plates by Zehner, lithographed by M. Fahrmbacher.

SCHOTT, HEINRICH WILHELM

AROIDEAE.

6 parts Folio (About 54 cm × 35 cm). Wien (Vienna) 1853 [–58].

60 plates by J. Oberer, J. Seboth and E. Nickelli, lithographed by J. Strohmeyer and N. Zehner.

SCHOTT, HEINRICH WILHELM

ICONES AROIDEARUM.

4 parts Folio (54 cm × 35 cm). Vindobonae (Vienna) 1857.

40 plates (pls. 1–30 hand-coloured) by J. Oberer, lithographed by M. Fahrmbacher, H. Sommer, J. Strohmayer.

SCHRADER, HEINRICH ADOLPH & WENDLAND, JOHANN CHRISTOPHER

SERTUM HANNOVERANUM SEU PLANTAE RARIORES QUAE IN HORTIS REGIIS HANNOVERAE VICINIS COLUNTUR.

4 Fasc. (in 1 vol.) Folio (46 cm × 29½ cm). (Parts 1–3), Goettingae (part 4). Hannoveae. 1795–98.

24 hand-coloured engraved plates all by J. C. Wendland.

Part 1, pp. 1–12, pls 1–6 (1795) with text by Schrader; parts 2–3, pp. 13–28, pls 7–18 (1796–97) by Schrader and Wendland; part 4, pp. 1–8, pls 19 (numbers 18–24) by Wendland. The work was continued in the same style and format by Wendland alone under the title *Hortus Herrenhusanus* (parts 1–4) in 1798–1801.

SCHRANK, FRANZ VON PAULA VON

PLANTAE RARIORES HORTI ACADEMICI MONACENSIS DESCRIPTAE ET OBSERVATIONIBUS ILLUSTRATE.

2 vols Folio (50 cm × 41 cm). Monachii (Munich) & Nürnberg (Nuremburg) [1817–] 1819 [–22].

100 unsigned lithographed plates.

According to W. T. Stearn in *J. Soc. Bibl. Nat. Hist.* 1: 151 (1938), Vol. 1, pls 1–20 published in 1817, pls 21–50 in 1819; vol. 2, pls 51–60 in 1820, pls 61–80 in 1821, pls 81–100 in 1822.

SCHULZE, MAX

DIE ORCHIDACEEN DEUTSCHLANDS, DEUTSCH-OESTERREICHS UND DER SCHWEIZ.

13 parts 8vo (26 cm × 17½ cm). Gera-Untermhaus [1892–] 1894.

95 chromolithographed plates (pls 1–70, 5b, 7b, 9b–c, 13b, 19b–c, 20b, 21b, 28b–c, 31b–d, 37b, 42b, 43b, 46b, 48b–d, 56b and duplicate pls 42, 65 and 69) and 1 uncoloured unnumbered plate by Bicknell, E. Fiek, M. Klee, J. Schulze and others.

PLATE 43

Priscilla Bury illustrated A Selection of Hexandrian Plants *(London 1831–34) with aquatint etchings. Printmaker Robert Havell concurrently etched the plates for Audubon's* The Birds of America, *and Audubon subscribed to Bury's beautiful book. She painted "Amaryllis Crocata" with its attendant butterfly "to preserve some memorial of the brilliant and fugitive beauties" of these species, as stated in the Preface.*

Gentiane sans tige.
Gentiana acaulis.
P.J. Redouté.
Langlois

SCHWEINFURTH, GEORG AUGUST

PLANTAE QUAEDAM NILOTICAE, QUAS . . . COLLEGIT ROBERTUS HARTMANN . . .
Folio (33½ cm × 25 cm). Berolini (Berlin) 1862.
16 plates by G. Schweinfurth and Hartmann, lithographed by G. Schweinfurth.

SCHWEINFURTH, GEORG AUGUST

RELIQUIAE KOTSCHYANAE. BESCHREIBUNG UND ABBILDUNG EINER ANZAHL . . . PFLANZENARTEN, WELCHE THEODOR KOTSCHY . . . IN DEN JAHREN 1837 BIS 1839 . . . IN DEN SÜDLICH VON KORDOFAN . . . GESAMMELT HAT.
4to (33 cm × 25 cm). Berlin 1868.
35 plates by Liepoldt, W. A. Meyn and J. Seboth, lithographed by W. A. Meyn, C. F. Schmidt and G. Schweinfurth.

SCOPOLI, GIOVANNI ANTONIO

DELICIAE FLORAE ET FAUNAE INSUBRICAE . . . QUAS IN INSUBRIA AUSTRIACA TAM SPONTANEAS, QUAM EXOTICAS VIDIT, DESCRIPSIT ET AERI INCIDI CURAVIT.
3 parts Folio (43 cm × 28 cm). Ticini (Pavia) 1786–88.
Title-page, frontispiece and 75 plates (of which 53 are botanical) by B. Bordiga, J. Cairoli, J. F. Chiesa and J. Lanfranchi, engraved by B. Bordiga and others.

SEEMANN, BERTHOLD CARL

THE BOTANY OF THE VOYAGE OF H.M.S. HERALD, UNDER THE COMMAND OF CAPT. HENRY KELLETT DURING THE YEARS 1845–51 . . .
4to (31 cm × 24½ cm). London 1852–57.
98 lithographed plates (pls 1–10, 13–100) by W. Fitch and J. D. Hooker, with hand-coloured frontispiece lithographed by Fitch from a sketch by T. Woodward, also 2 maps probably representing Plates 11 and 12.
For dates of publication, see T. A. Sprague in *J. Bot. (London)* 59:22 (1921); *Flora Malesiana* I. 4: ccxii (1954).

SEEMANN, BERTHOLD CARL

FLORA VITIENSIS: A DESCRIPTION OF THE PLANTS OF THE VITI OR FIJI ISLANDS.
4to (31 cm × 24½ cm). London 1865–73.
99 hand-coloured plates mostly by W. Fitch but pl. 1 by Macdonald, 2 plates by Mitten, some analyses by Solms-Laubach and H. Wendl, lithographed by W. Fitch.

SIBTHORP, JOHN & SMITH, JAMES EDWARD

FLORA GRAECA; SIVE PLANTARUM RARIORUM HISTORIA, QUAS IN PROVINCIIS AUT INSULIS GRAECIAE LEGIT, INVESTIGAVIT ET DEPINGI CURAVIT. J. SIBTHORP CHARACTERES OMNIUM DESCRIPTIONES ET SYNONYMA ELABORAVIT J. E. SMITH (VOLS 1–7), J. LINDLEY (VOLS 8–10).
10 vols Folio (47 cm × 32 cm). London 1806–40.
966 hand-coloured plates by Ferdinand Bauer, engraved by Ferd. Bauer, James Sowerby and J. de C. Sowerby.
One of the rarest and most beautiful in both plates and typography of all botanical works. Only 25 complete sets of the original issue were published. Sibthorp planned the work but did not live to write it. Smith undertook the writing of descriptions and at his death in 1828 had prepared seven volumes; Lindley completed the work. The entire cost of the undertaking exceeded £30,000; thus each set issued at £254 cost more than £1,000 to produce. Between 1845 and 1856 Henry G. Bohn issued a new impression of about 40 copies, slightly inferior in colouring to the original and recognisable by the watermark '1845' on many plates.

SINCLAIR, ISABELLA (MRS FRANCIS SINCLAIR)

INDIGENOUS FLOWERS OF THE HAWAIIAN ISLANDS.
Folio (36 cm × 26 cm). London 1887.
44 chromolithographed plates by I. Sinclair.
According to H. St. John in *Pacific Science* 8: 140–146 (1954) this is 'the first book with colour pictures of Hawaiian flowering plants . . . Even today it has more colour plates of Hawaiian plants than any other book'. St. John, *loc. cit.*, provides modern botanical names and comments on the plants figured.

SINNING, WILHELM. See NEES VON ESENBECK, T.F.L. 1825.

SLOANE, SIR HANS

A VOYAGE TO THE ISLANDS MADEIRA, BARBADOS, NIEVES, S. CHRISTOPHERS AND JAMAICA, WITH THE NATURAL HISTORY OF THE LAST OF THOSE ISLANDS . . .
2 vols Folio (34½ cm × 26 cm). London 1707–25.
285 uncoloured plates (vol. 1 pls. I–IV, 1–155, vol. 2 pls. V–XI, 157–274), engraved by M. v. d. Gucht and J. Savage.

PLATE 44

"Gentiane sans tige" illustrates Pierre Joseph Redouté's Choix des plus belles fleurs *(Paris 1827–33). The plate title is a misnomer, as this flower is a peony, and later prints pulled from this plate show the corrected title, "Pivoine officinale à fleurs simples." For details about the book we photographed for this and the following plate, see the List of the Plates.*

Commonly cited as Sloane's *Natural History of Jamaica* or *Hist. Jam.*, this is a fundamental work for West Indian botany. The botanical plates, although of little artistic merit, are scientifically important, being accurate representations 'as big as the Life' of the dried specimens in Sloane's herbarium (at the British Museum (Nat. Hist.)) which has the original drawings mounted alongside them. Many of these have the inscription 'Everhardus Kickius fecit' or 'E. Kickius fecit', and he would appear to have drawn them all.

SMITH, F. W.
THE FLORIST'S MUSEUM, A REGISTER OF THE NEWEST AND MOST BEAUTIFUL VARIETIES OF FLORIST'S FLOWERS
4to. London n.d. (1835).
Coloured title-page and 59 hand-coloured lithographs.
Also published with title-page *Florist's Magazine* 1836.
H.B.

SMITH, SIR JAMES EDWARD
PLANTARUM ICONES HACTENUS INEDITAE, PLERUMQUE AD PLANTAS IN HERBARIO LINNAEANO CONSERVATAS DELINEATAE.
3 parts Folio (37½ cm × 23½ cm). London 1789–91.
75 unsigned hand-coloured engraved plates all by J. Sowerby.

SMITH, SIR JAMES EDWARD
ICONES PICTAE PLANTARUM RARIORUM, DESCRIPTIONIBUS ET OBSERVATIONIBUS ILLUSTRATAE . . .
3 parts Folio (60 cm × 48 cm). London 1790 [–93].
18 hand-coloured engraved plates all by J. Sowerby.
Part 1, pls 1–6, Oct., 1790; part 2, pls 7–12, Feb. 1792; part 3, pls 13–18, Nov. 1793.

SMITH, SIR JAMES EDWARD
SPICILEGIUM BOTANICUM. GLEANINGS OF BOTANY.
2 parts Folio (32 cm × 20 cm). London 1791–92.
24 unsigned hand-coloured engraved plates all by J. Sowerby.

SMITH, SIR JAMES EDWARD
A SPECIMEN OF THE BOTANY OF NEW HOLLAND.
1 vol. (in 4 parts). 4to (30 cm × 23 cm). London 1793 (–95).
16 hand-coloured engraved plates all by J. Sowerby.

SMITH, SIR JAMES EDWARD
EXOTIC BOTANY: CONSISTING OF COLOURED FIGURES, AND SCIENTIFIC DESCRIPTIONS OF SUCH NEW, BEAUTIFUL, OR RARE PLANTS AS ARE WORTHY OF CULTIVATION IN THE GARDENS OF BRITAIN . . . THE FIGURES BY JAMES SOWERBY.
2 vols 4to (29 cm × 22½ cm), 8vo (23 cm × 13½ cm). London 1804–05 (–1808).
120 hand-coloured engraved plates all by J. Sowerby.

SMITH, SIR JAMES EDWARD & JOHN ABBOT
NATURAL HISTORY OF THE LEPIDOPTEROUS INSECTS OF GEORGIA . . . AND THE PLANTS ON WHICH THEY FEED.
Folio (40 cm × 31 cm). London 1797.
104 hand-coloured engravings. H.B.

SMITH, MISS
STUDIES OF FLOWERS FROM NATURE . . .
4to (38 cm × 27 cm).
Printed for and sold by Miss Smith, Adwick Hall, near Doncaster. n.d. (1820).
Coloured title-page and twenty plates in two states, uncoloured and hand-coloured aquatints. H.B.

SOLANDER, DANIEL CARL. See BANKS, SIR JOSEPH. 1900

SOWERBY, JAMES
COLOURED FIGURES OF ENGLISH FUNGI OR MUSHROOMS.
Folio London [1795–] 1797–1803 [–1815]
440 hand coloured and colour-printed engraved plates on 436 leaves, by James Sowerby.

SOWERBY, JAMES
FLORA LUXURIANS OR THE FLORIST'S DELIGHT
3 parts Folio. London 1789–1791.
18 hand-coloured engravings. No title-page.
See also W. Blunt, *The Art of Botanical Illustration*.
Very rare. H.B.

SOWERBY, JAMES & SMITH, SIR JAMES EDWARD
ENGLISH BOTANY, OR, COLOURED FIGURES OF BRITISH PLANTS WITH THEIR ESSENTIAL CHARACTERS, SYNONYMS, AND PLACES OF GROWTH. TO WHICH WILL BE ADDED OCCASIONAL REMARKS.
36 vols 8vo (24 cm × 15 cm). London 1790–1814.
2592 hand-coloured engraved plates by James Sowerby.
Title-page to vol. 1 does not bear Smith's name.
Supplement (by W. J. Hooker, W. Borrer, C. C. Babington, M. J. Berkeley, W. Wilson and others).
5 vols 8vo (25 cm × 15½ cm). London 1831–66.
407 hand-coloured engraved plates (pls 2593–2999) by J. de C. Sowerby and J. W. Salter.
Ed. 2. Edited by C. Johnston. 12 vols 8vo (21 cm × 13 cm). London 1832–46.
'A cheap and humble edition; the plates only partly coloured'.
Ed. 3. Edited by J. T. Boswell Syme.

12 vols 8vo (26 cm × 17½ cm). London 1863–86.
1922 hand-coloured lithographed plates copied from the earlier engraved plates by J. E. Sowerby, but inferior, but with others added by N. E. Brown, and W. H. Fitch. Pl. 1834* by N. E. Brown (of *Lycopodium alpinum* var. *decipiens* and var. *anceps*) is in a few copies.
Ed. 3 vol. 13 (Supplement to Third Edition vols 1–4) by N. E. Brown.
8vo (25 cm × 17 cm). London 1891–92.
14 hand-coloured lithographed plates by W. H. Fitch.
Although part 1 is dated '1891' all 3 parts appeared in 1892.

SPACH, EDOUARD
HISTOIRE NATURELLE DES VÉGÉTAUX. PHANÉROGAMES.
14 vols & Atlas. 8vo (23 cm × 15½ cm). Paris 1834–48.
152 hand-coloured plates by J. Decaisne, F. Legendre & Mme Spach, engraved by Plée, Breton, Davesne, Mlle Dondet, Le Couturier, F. Legendre, Mougot and Mlle Noiret.

SPACH, EDOUARD. See JAUBERT, HIPPOLYTE FRANÇOIS 1842.

SPAËNDONCK, GERRIT VAN
FLEURS DESSINÉES D'APRÈS NATURE.
Folio (49 cm × 33 cm). Paris [1801].
24 unnumbered plates (partly colour-printed) by G. v. Spaëndonck, engraved by P. F. Le Grand, Carrée, A.L.X. Chaponnier and I. Godefroy.

SPAËNDONCK, GERRIT VAN
SOUVENIRS DE VAN SPAËNDONCK, OU RECUEIL DE FLEURS, LITHOGRAPHIÉES D'APRÈS LES DESSINS DE CE CÉLÈBRE PROFESSEUR . . .
4to (22½ cm × 26½ cm). Paris 1826.
19 hand-coloured plates by Spaëndonck, lithographed by G. Engelmann.

STERLER, ALOIS & MAYRHOFFER, JOHANN NEPOMUCK
EUROPA'S MEDIZINISCHE FLORA.
20 parts Folio (39 cm × 25 cm). Munich 1820.
80 hand-coloured engravings. H.B.

STERNBERG, GASPAR COMTE DE
REVISIO SAXIFRAGARUM ICONIBUS ILLUSTRATA with SUPPLEMENTUM and SUPPLEMENTUM SECUNDUM.
Large Folio (47 cm × 30 cm). Ratisbonae, 1810, 1822; Prague 1831.
57 coloured and uncoloured engravings.
1st part published at Regensburg 1810, 1st supp. Regensburg 1822.
2nd supplement Prague 1831. H.B.

STURM, JACOB
DEUTSCHLANDS FLORA, IN ABBILDUNGEN NACH DER NATUR MIT BESCHREIBUNGEN.
163 parts small 8vo (12½ cm × 8½ cm). Nürnberg (Nuremberg) 1798–1862.
2472 engraved plates, chiefly by Jacob and J. W. Sturm.
Plates very small, but charming and detailed.

SWARTZ, OLOF
ICONES PLANTARUM INCOGNITARUM, QUAS IN INDIA OCCIDENTALI DETEXIT ATQUE DELINEAVIT.
Folio (39½ cm × 26 cm). Erlangae (Erlangen) 1794–1800.
13 hand-coloured plates by O. Swartz, engraved by J. F. Volkart.
Sect. 1, pp. 1–8, pls 1–6 in 1794, sect. 2, pls 7–13 in 1800; pls 14–25, listed in table of contents to Fasciculus 1, never published.
Apparently Swartz made about 200 drawings for this work, which he sent to Erlangen. About 1890 Urban acquired 71 of them, which are cited by him and others as 'Sw. Ico. ined.' (c.f. Urban, *Symbolae Antillanae* 1: 164; 1898). In March 1943 these were destroyed at Berlin by bombing.

SWEET, ROBERT
GERANIACEAE. THE NATURAL ORDER OF GERANIA, ILLUSTRATED BY COLOURED FIGURES AND DESCRIPTIONS . . .
5 vols 8vo (23 cm × 14½ cm). London 1820–1830.
500 hand-coloured plates (pls 1–400, 1–100) by E. D. Smith and M. Hart, engraved by S. Watts.

SWEET, ROBERT
THE BRITISH FLOWER-GARDEN; CONTAINING COLOURED FIGURES AND DESCRIPTIONS OF . . . HERBACEOUS PLANTS . . .
3 vols 8vo (23 cm × 14 cm). London 1823–29.
300 hand-coloured plates by E. D. Smith, engraved by J. Alais, A. Baily and Weddell.
Ser 2. 4 vols 8vo (23 cm × 14 cm). London [1829–] 1831–38.
412 hand-coloured plates by E. D. Smith, F. W. Smith, T. Allport, J. Hart, J. Humble Eden, R. Kippist, J. Macnab and Miss Selwyn, engraved by Weddell, F. W. Smith and Barclay.
2nd ed. 1838.

SWEET, ROBERT
CISTINEAE. THE NATURAL ORDER OF CISTUS, OR ROCK-ROSE; ILLUSTRATED BY COLOURED FIGURES AND DESCRIPTIONS . . .
8vo (24 cm × 15 cm). London 1825–30.
112 hand-coloured engraved plates by J. Hart, M. Hart, W. Hart and Mrs Brown, engraved by S. Watts and Weddell.

Hortensia

P. J. Redouté Langlois.

SWEET, ROBERT

FLORA AUSTRALASICA; OR A SELECTION OF HANDSOME OR CURIOUS PLANTS, NATIVES OF NEW HOLLAND AND THE SOUTH-SEA-ISLANDS, CONTAINING COLOURED FIGURES AND DESCRIPTIONS . . .
8vo (23½ cm × 15 cm). London 1827–28.
56 hand-coloured plates by E. D. Smith, engraved by S. Watts.

SWEET, ROBERT

THE FLORIST'S GUIDE AND CULTIVATOR'S DIRECTORY; CONTAINING COLOURED FIGURES . . .
2 vols 8vo (25 cm × 15½ cm). London 1827–32.
200 hand-coloured plates by E. D. Smith, E. Hogg, W. Prest and I. Thackery, engraved by S. Watts.

TARGIONI-TOZZETTI, ANTONIO

RACCOLTA DI FIORI, FRUTTI ED AGRUMI . . .
14 parts Folio (46½ cm × 32 cm). Firenze (Florence) [1822–] 1825 [–1830].
42 colour-printed and partly hand-coloured unnumbered plates by G. Angiolini, G. Geri, F. Mariti, A. Maryes, P. Morini, (Murini), O. Muzzi, P. Tofani, engraved by G. Pera, B. Rosaspina and D. Serantoni.

TAUSCH, IGNAZ FRIEDRICH

HORTUS CANALIUS SEU PLANTARUM RARIORUM QUAE IN HORTO BOTANICO . . . J. MALABAILA COMITIS DE CANAL COLUNTUR, ICONES ET DESCRIPTIONES.
1 vol. (2 parts). Folio (part I 48 cm × 34½ cm; part II 49 cm × 35 cm). Pragae (Prague) 1823.
20 unnumbered lithographed plates (hand-coloured in Dec. 1 only) by F. Both and Kraupa, lithographed by A. Gareis, Kraupa, W. Krüger, C. Stafferhagen, C. Zimmermann.
See F. Widder, 'Actaea nutans Tausch und der Hortus Canalius', *Phyton* 1: 258–268 (1949).

TENORE, MICHELE

FLORA NAPOLITANA, OSSIA DESCRIZIONE DELLA PIANTE INDIGENE DEL REGNO DI NAPOLI . . .
5 vols Folio (44 cm × 31 cm), and Atlas, Folio (56½ cm × 40 cm). Napoli 1811–38.
250 hand-coloured plates by Federigo Dehnhardt, Giuseppe Lettieri and Steurnal, Carlo and R. Biondi and others.

THORNTON, ROBERT JOHN

NEW ILLUSTRATION OF THE SEXUAL SYSTEM OF CAROLUS VON LINNAEUS . . .
Folio (56 cm × 44 cm). London (1799–) 1807.
This contains *The Temple of Flora* (see below) and 'an elucidation of the several parts of the fructification, a prize dissertation on the sexes of plants' etc. together with 25 engraved portraits and 60 to 80 engraved plates.

THORNTON, ROBERT JOHN

THE TEMPLE OF FLORA, OR GARDEN OF NATURE, BEING PICTURESQUE BOTANICAL PLATES OF THE NEW ILLUSTRATION OF THE SEXUAL SYSTEM OF LINNAEUS . . .
Folio (56 cm × 44 cm). London (1799–) 1807.
31 coloured engraved plates by Henderson, Reinagle, Bishop, S. Edwards, Ehret, Miller, Pether, Sowerby, Sang-so, Thornton, Wo-sion, engraved by many artists.
As stated by Buchanan, 'no two copies of this book are alike . . . The plates were engraved by various processes—aquatint, mezzotint, stipple and line engraving—and the impressions taken from them were printed in colour and finished by hand . . . most of the plates were altered or added to from time to time, producing a different 'state' in each case, some plates having as many as four different states'. For further information about these, reference must be made to G. Grigson and H. Buchanan, *Thornton's Temple of Flora* (1951).
[Lottery edition:] TEMPLE OF FLORA, OR GARDEN OF THE BOTANIST, POET, PAINTER AND PHILOSOPHER.
Folio (37 cm × 31 cm). London 1812.
32 plates, much inferior to those of large folio editions.

TIRPENNE

MÉTHODE TIRPENNE.
5 vols Folio. Paris 1840.
The fifth volume contains: Cours de Fleurs sous la Direction de M. Redouté par M. A. Prévost. 5 uncoloured plates each with 5 drawing stages in red after Redouté. Cours de Fleurs par M. A. Prévost, d'après M. Redouté: 1 uncoloured plate with 5 drawing stages in red.

PLATE 45

"Hortensia" illustrates Choix des plus belles fleurs *(Paris 1827–33) by Pierre Joseph Redouté. His use of color-printing permits these delicate petals to glow with exceptional realism. Redouté's* Choix *often survives with darkened paper, but this hydrangea comes from a special issue of the book that is as crisp as the day it was printed.*

TITFORD, WILLIAM JOWIT

SKETCHES TOWARDS A HORTUS BOTANICUS AMERICANUS; OR, COLOURED PLATES . . . OF NEW AND VALUABLE PLANTS OF THE WEST INDIES AND NORTH AND SOUTH AMERICA . . .
4to (29 cm × 23 cm). London 1811–12.
Frontispiece and 17 hand-coloured plates by W. J. Titford, engraved by W. N.

TODARO, AGOSTINO

HORTUS BOTANICUS PANORMITANUS, SIVE PLANTAE NOVAE VEL CRITICAE QUAE IN HORTO BOTANICO PANORMITANO COLUNTUR DESCRIPTAE ET ICONIBUS ILLUSTRATAE.
2 vols Folio (44 cm × 31½ cm). Panormi (Palermo) 1875–92.
40 chromolithographed plates by A. Ficarrotta, lithographed by C. Visconti and A. and N. Maniaci.

TRAILL, CATHERINE PARR

CANADIAN WILD FLOWERS.
Folio (34 cm × 27 cm). Montreal 1869.
Painted and lithographed by Agnes Fitzgibbon.
Coloured title-page and 10 hand-coloured lithographs. H.B.

TRATTINICK, LEOPOLD

THESAURUS BOTANICUS.
20 parts Folio (38 cm × 25 cm). Viennae (1805–) 1819.
80 hand-coloured plates by Strenzel, Frister, Kneisech, J. Schmid, Reinelli, Malknecht and Buchberger, engraved by Belling, C. Hanauer, A. Sommer and Weber.

TRATTINICK, LEOPOLD

AUSWAHL VORZÜGLICH SCHÖNER, SELTENER, BERÜHMTER, UND SONST SEHR MERKWÜRDIGER GARTENPFLANZEN, IN GETREUEN ABBILDUNGEN . . .
2 vols 4to (25½ cm × 17½ cm). Wien (Vienna) 1821.
218 unnumbered hand-coloured engraved plates by Baron Eyb and Strenzel, engraved by L. Rieder and J. Seher. 2 engraved portraits.

TRATTINICK, LEOPOLD

NEUE ARTEN VON PELARGONIEN DEUTSCHEN URSPRUNGES. ALS BEYTRAG ZU ROB. SWEET'S GERANIACEEN MIT ABBILDUNGEN UND BESCHREIBUNGEN.
6 vols 8vo (24½ cm × 16½ cm). Wien (Vienna) 1825–43.
202 hand-coloured engraved plates (with 264 figures) by Fr. Blaschek, Parger and Seiler, engraved by Lor. Rieder.

TREW, CHRISTOPH JAKOB

PLANTAE SELECTAE, QUARUM IMAGINES . . . PINXIT G. D. EHRET . . . COLLEGIT J. C. TREW . . . [DEC. 1–7] POST IPSIUS MORTEM NOMINIBUS ET NOTIS ILLUSTRAVIT . . . B. C. VOGEL (DEC. 8–10) . . . IN AES INCIDIT ET VIVIS COLORIBUS . . . J. J. HAID.
10 parts Folio (54 cm × 37 cm). Augustae Vind. (Augsburg) 1750–73.
100 hand-coloured plates by G. D. Ehret, engraved by J. J. and E. D. Haid.
SUPPLEMENTUM PLANTARUM SELECTARUM . . .
Dec. 1 and 2. Folio (45 cm × 37 cm). Augustae Vind. (Augsburg). 1790–92.
20 hand-coloured plates (Nos 101–120) by G. D. Ehret, engraved by J. E. Haid.
Many fine plates. W.B.

TREW, CHRISTOPHER JAKOB

HORTUS NITIDISSIMIS OMNEM PER ANNUM SUPERBIENS FLORIBUS, SIVE AMOENISSIMORUM FLORUM IMAGINES, QUAS MAGNIS SUMPTIBUS COLLEGIT, IPSO VERO ANNUENTE IN AES INCISAS VIVISQUE COLORIBUS PICTAS, IN PUBLICEM EDIDIT J. M. SELIGMANN (Vols. 2–3, A.L. WIRSING).
3 vols Folio (50½ cm × 34 cm). Norimbergae (Nuremberg) [1750–] 1786.
180 hand-coloured plates by G. D. Ehret, G. W. Baurenfeind, B. R. and M. B. Dietzsch, Eisenberger, M. M. Heumannin, Karell, J. C. Keller, A. I. Meyer, M. M. Payerlein, Siverts, C.J.C. Wirsing, engraved by J. M. Seligmann, J. M. Stock, A. L. Wirsing.
Plates 60–61 and 121–122 are represented by one plate each.

TREW, CHRISTOPH JAKOB

PLANTAE RARIORES, QUAS MAXIMAM PARTEM IPSE IN HORTO DOMESTICO COLUIT . . . EDENTE J. C. KELLER.
Folio Norimbergae (Nuremberg) Dec. I 1763, Dec. II and III . . . CURAM ET ILLUSTRATIONEM SUSCEPIT B.C. VOGEL . . . ET A.L. WIRSING, 1799–84.
30 hand-coloured plates by C. W. Baurenfeind, G. D. Ehret, N. F. Eisenberger, M. M. Payerlein, engraved by J. C. Keller & A. L. Wirsing.

TRIANA, JÉRONIMO

NOUVELLES ÉTUDES SUR LES QUINQUINAS . . . ACCOMPAGNÉES DE FACSIMILE DES DESSINS DE LA QUINOLOGIE DE MUTIS . . .
Folio (51½ cm × 35 cm). Paris 1870 [actually 1872].
33 hand-coloured plates (pls. 2–31, 20 bis, and two plates each numbered 22 and 24 made from photographs of drawings by Mutis) lithographed by E. Grabowski. There is no pl. 1. The title-page is dated 1870 but owing to the Franco-German War publication did not take place until 8th April 1872.

TRIMEN, HENRY. See BENTLEY, ROBERT. 1880.

TRINIUS, KARL BERNHARD
SPECIES GRAMINUM ICONIBUS ET DESCRIPTIONIBUS ILLUSTRAVIT.
3 vols 4to (28½ cm × 19 cm). Petropoli (St Petersburg); Lipsiae (Leipzig) [1825–] 1828–36.
360 uncoloured lithographed plates by W. Pape.

TRINIUS, KARL BERNHARD. See LIBOSCHITZ, JOSEPH, 1811.

TURPIN, PIERRE JEAN FRANÇOIS
ESSAI D'UNE ICONOGRAPHIE ÉLÉMENTAIRE ET PHILOSOPHIQUE DES VÉGÉTAUX, AVEC UN TEXTE EXPLICATIF.
8vo (21½ cm × 13½ cm) approx. Paris 1820.
64 hand-coloured plates (pls 1–56, 2 bis, 4 bis, suite de 4 bis, 36 bis, 43 bis, 44 bis, 48 bis, 56 bis) and 1 large uncoloured folded-in engraved plate ('Organographie Végétal') by Turpin, engraved by Plée, Rebel, Massard, Guyard, Boutelou and others.
Issued also as Vol. 3 of J.L.M. Poiret, *Leçons de Flore*, and Vol. 8 of Chaumeton, *Flore Médicale*.

TUSSAC, F. RICHARD DE
FLORE DES ANTILLES . . . DÉCRITS D'APRÈS NATURE . . . AVEC ['ENRICHIES DE'] PLANCHES . . . COLORIÉES.
4 vols Folio (53 cm × 35 cm). Paris 1808–27.
140 hand-coloured plates (pls 1–30, 25 bis; 1–24 and 10 unnumbered; 1–37, 1; 1–37) by Turpin, Poiteau, P. J. Redouté and others, engraved by Dien, Gabriel, Massard, Giraud, Bouquet and others.

TWINING, ELIZABETH
ILLUSTRATIONS OF THE NATURAL ORDERS OF PLANTS. ARRANGED IN GROUPS; WITH DESCRIPTIONS.
2 vols Folio (50 cm × 33½ cm). London 1849–55.
160 hand-coloured lithographed plates all by E. Twining.
Another edition (reduced from original folio) 8vo (24½ cm × 15 cm). London 1868.
Plates colour-printed.

TYAS, ROBERT
FAVOURITE FIELD FLOWERS; OR, WILD FLOWERS OF ENGLAND POPULARLY DESCRIBED . . .
2 series in 8vo (16½ cm × 10 cm). London [1847–] 1848–50.
24 hand-coloured lithographed plates (12 in each series) by James A. Andrews.
Plates have much charm and are beautifully executed, but are on a very small scale.

TYAS, ROBERT
FLOWERS FROM THE HOLY LAND . . .
8vo (16½ cm × 10 cm). London (1850–) 1851.
12 hand-coloured lithographed plates by J. Andrews.

TYAS, ROBERT
FLOWERS FROM FOREIGN LANDS; THEIR HISTORY AND BOTANY, WITH CONCISE DESCRIPTIONS OF THEIR NATIVE REGIONS.
8vo (16½ cm × 10½ cm). London 1853.
12 hand-coloured lithographed plates by James Andrews.

VAILLANT, SÉBASTIEN
BOTANICON PARISIENSE, OU DÉNOMBREMENT . . . DES PLANTES QUI SE TROUVENT AUX ENVIRONS DE PARIS.
Folio (40½ cm × 25½ cm). Leide (Leyden) & Amsterdam 1727.
33 plates by C. Aubriet engraved by Wandelaar.

VAN GEEL, PIERRE CORNEILLE
SERTUM BOTANICUM. COLLECTION DE PLANTES REMARQUABLES PAR LEUR UTILITÉ.
Folio. Bruxelles 1828–1832.
600 hand-coloured lithographs.
This work was presumably pirated since most if not all of the plates are taken from other books. Some from Curtis's *Botanical Magazine*; some are the work of Sydenham Edwards; some of Andrews and some of Redouté. (Included here only to complete the record.)

VELLOSO, JOSÉ MARIANNO DA CONCEIÇAO
. . . FLORAE FLUMINENSIS ICONES NUNC PRIMO EDUNTUR. EDITIT A. DE ARRABIDA.
11 vols Folio (51½ cm × 33 cm). Parisiis (Paris) 1827 [actually 1835].
1640 plates.
Vol. 1, pls 1–153; Vol. 2, pls 1–156; Vol. 3, pls 1–168; Vol. 4, pls 1–189; Vol. 5, pls 1–135; Vol. 6, pls 1–113; Vol. 7, pls 1–164; Vol. 8, pls 1–164; Vol. 9, pls 1–128; Vol. 10, pls 1–143; Vol. 11, pls 1–127) by Velloso and F. Solano, lithographed by J. and E. Knecht.
This work had a very unfortunate history. Ten years after the death of Velloso (1743–1812) the Emperor Pedro I of Brazil decided that Velloso's large collection of drawings, mostly bold outlines, of Brazil plants growing near Rio de Janeiro (Sebastianopolis) should be published. They were sent to Paris and lithographed in an impression of 3,000 copies but not issued before the abdication of Pedro I in 1831 rendered them worthless and the whole stock, except for 100 copies, was treated as waste paper. According to De Candolle, quoted by O. Kuntze, *Rev. Gen. Pl.* 1: cxlv (1891), the remaining copies were published in 1835, evidently late in the year, for they were first listed for sale in *Bibl. France* 1836 (3): 34 (16 Jan. 1836).
In matters of detail these plates leave much to be desired.

VENTENAT, ÉTIENNE PIERRE
DESCRIPTION DES PLANTES NOUVELLES ET PEU CONNUES, CULTIVÉES DANS LE JARDIN DE J. M. CELS. AVEC FIGURES.

MONS ATHOS AB OCCIDENTE.

10 parts Folio (46 cm × 30 cm). Paris An VIII [i.e. 1800–03].
100 uncoloured plates by P. J. Redouté, Cloquet, Laneau, Maréchal and Sauvage, engraved by S. Goulet, J. B. Guyard, Plée père et fils, and Sellier.
For dates of publication see W. T. Stearn in *J. Soc. Bibl. Nat. Hist.* 1: 199 (1939).

VENTENAT, ÉTIENNE PIERRE
JARDIN DE LA MALMAISON.
2 vols Folio (15½ cm × 35½ cm). Paris An XI–XII (1803–4) [i.e. 1803–1805].
120 stipple-engravings printed in colour and finished by hand, by P. J. Redouté, engraved by P. F. Legrand, J. B. Dien, L. J. Allais and others.
For dates of publication see W. T. Stearn in *J. Soc. Bibl. Nat. Hist.* 1: 200 (1939).

VENTENAT, ÉTIENNE PIERRE
CHOIX DE PLANTES, DONT LA PLUPART SONT CULTIVÉES DANS LE JARDIN DE M. CELS.
10 parts Folio (51 cm × 33½ cm). Paris An XI 1803 [–08].
60 colour-printed plates by P. Bessa, Poiteau, P. J. Redouté and Turpin, engraved by Dien, Plée and Sellier.
For dates of publication see A. W. Exell in *J. Bot. (London)* 76: 181 (1936).
W. T. Stearn in *J. Soc. Bibl. Nat. Hist.* 1: 201 (1939).

VIETZ, FERDINAND BERNHARD
ICONES PLANTARUM MEDICO-OECONOMICO-TECHNOLOGICARUM CUM EARUM FRUCTUS USUSQUE DESCRIPTIONE ODER ABBILDUNGEN ALLER MEDIZINISCH-ÖKONOMISCH-TECHNOLOGISCHEN GEWÄCHSE . . .
10 vols 4to (24 cm × 18½ cm). Wien (Vienna) (1800–) 1819 [–20].
1086 hand-coloured engraved plates (pls. 1–935, with 151 additional plates) by I. Albrecht. The 151 additional plates are all called 'a', 'b', 'c' etc. as follows:
Vol.
2: 2 plates to nos 114, 172, 198.
3: 2 plates to nos 226, 230, 234, 237, 239, 248, 250, 251, 254, 261, 263, 265, 267;
3 plates to nos 227, 231, 233, 252, 253, 259, 260, 264, 266, 268, 271;
4 plates to no. 241;
19 plates to no. 270.
4: 2 plates to nos 277, 296, 316, 319, 320, 321, 327, 331, 333, 355;
3 plates to nos 296, 300, 344, 345;
4 plates to no. 313.
5: 2 plates to nos 372, 376, 382, 389, 391, 400, 404, 416, 423, 427, 433, 437, 439, 441, 450;
4 plates to no. 403;
5 plates to no. 440.
6: 2 plates to nos 464, 474, 476, 483, 491, 496, 510, 511, 534–36, 539;
3 plates to no. 477;
5 plates to no. 481.
7: 2 plates to nos 553, 561, 577, 585, 593, 596, 599–609, 610, 613, 614, 616, 625;
3 plates to nos 556, 579, 583;
4 plates to no. 611.
8: 2 plates to nos 653, 663, 667, 677, 690, 706, 710, 712.
10: 2 plates to no. 854.
Supplement (Vol. II) by J. N. Kerndl. Wien (Vienna) 1822.
100 hand-coloured engraved plates by I. Albrecht.

VIGNEUX, A.
FLORE PITTORESQUE DES ENVIRONS DE PARIS.
4to (27 cm × 20 cm). Paris 1812.
62 hand-coloured engravings. H.B.

VINCENT, MADAME HENRIETTE ANTOINETTE
ÉTUDES DE FLEURS ET DE FRUITS PEINTS D'APRÈS NATURE.
Folio. Paris n.d. [1810].
48 plates printed in colour and finished by hand.
A very fine work. H.B.
English Editions of the above work.
Dunthorne has clearly gone astray for once in his collations of these. It is therefore only possible to give below particulars of two books listed by him (320 and 321) which have been personally inspected, but differ from his descriptions in that those given below possess more plates.

VINCENT, MADAME HENRIETTE ANTOINETTE
THE ELEMENTS OF FRUIT AND FLOWER PAINTING.
Folio (31 cm × 27 cm). London (R. Ackermann) 1814.
Illustrated with engravings by T. L. Bushby from studies after nature by Madame Vincent.
23 plates in two states uncoloured and printed in colours and finished by hand.

PLATE 46
Three editors carried John Sibthorp's Flora graeca *(London 1806–40) to completion over thirty-five years, yet a grand consistency was upheld throughout. Ferdinand Bauer drew all the plates, and the artist, with James Sowerby and his son, etched its 966 plates. Each of the ten volumes has a beautiful title page with a vignette such as this of Mount Athos. Even the calligraphic designer, Tomkins, is credited with a line at lower left.*

VINCENT, MADAME HENRIETTE ANTOINETTE

STUDIES FROM FRUIT AND FLOWERS PAINTED FROM NATURE AND ENGRAVED BY T. L. BUSHBY.

Folio (31 cm × 27 cm). London (R. Ackermann) 1814.

18 plates in two states uncoloured and printed in colours by B. M'Queen. H.B.

VISIANI, ROBERTO DE

FLORA DALMATICA; SIVE, ENUMERATIO STIRPIUM VASCULARIUM QUAS HACTENUS IN DALMATIA LECTAS ET SIBI OBSERVATAS DESCRIPSIT, DIGESSIT RARIORUMQUE ICONIBUS ILLUSTRAVIT.

3 vols 4to (27½ cm × 21½ cm). Lipsiae (Leipzig) 1842–52.

57 hand-coloured plates (pls 1–55, 10bis and 10ter) by J. Clementi, engraved by G. Alboth, A. Harzer, G. Langer, C. Schnorr, A. Weidenbach and G. Zumpe. Supplementum (in *Mem. Ist. Veneto*, 16 1872), 10 plates.

Supplementum II (*Loc. cit.* 20, 21, 1877–82), 8 plates.

VOLCKAMER, JOHANN CHRISTOPH

NÜRNBERGISCHE HESPERIDES, ODER GRÜNDLICHE BESCHREIBUNG DER EDLEN CITRONAT-, CITRONEN- UND POMERANTZEN-FRÜCHTE . . .

Folio (34 cm × 21½ cm). Nürnberg (Nuremberg) 1708.

116 plates and 18 vignettes by P. Decker (and J. C. and J. G. Volckamer?) engraved by P. Decker, L. C. Glotsch, B. Kenckel, C. F. Krieger, J. A. Montalegre, W. Pfan and J. C. Steinberger.

An attractive book with decorative plates of fruit, landscape and houses. W.B.

VRIESE, WILLEM HENDRIK DE

DESCRIPTIONS ET FIGURES DES PLANTES NOUVELLES ET RARES DU JARDIN BOTANIQUE DE L'UNIVERSITÉ DE LEIDE . . .

2 parts Folio (50 cm × 36 cm). Leide (Leyden) & Leipzig 1847 [–51].

10 hand-coloured plates by P. W. M. Trap, C. F. Brede Jn. and E. J. Koning, lithographed by J. P. Berghaus and A. J. Wendel.

VRIESE, WILLEM HENDRIK DE

GOODENOVIEAE . . .

4to (29 cm × 23 cm). Harlemi (Haarlem) 1854.

38 lithographed plates (one hand-coloured) by Q.M.R. Ver Huell and A. J. Wendel.

Published as a separate from *Natuurk. Verhandel. Hollandsche Maatsch. Wetensch. te Haarlem* 10 (1854).

VRIESE, WILLEM HENDRIK DE

ILLUSTRATIONS D'ORCHIDÉES DES INDES ORIENTALES NÉDERLANDAISES, OU CHOIX DE PLANTES NOUVELLES ET PEU CONNUES, DE LA FAMILLE DES ORCHIDÉES.

3 parts (plates 70 cm × 52 cm, folded as 52 cm × 35 cm). La Haye (The Hague) 1854.

18 plates (17 chromolithographed) by J. v. Aken, lithographed by C. W. Mieling.

VRIESE, WILLEM HENDRIK DE

ILLUSTRATIONS DES RAFFLESIAS ROCHUSSENII ET PATMA, D'APRÈS LES RECHERCHES FAITES AUX ÎLES DE JAVA ET DE NOESSA KAMBANGAN PAR J. E. TEYSMANN ET S. BINNENDYK.

Folio (Text, 47½ cm × 30½ cm, plates, 72 cm × 50 cm and 50 cm × 36 cm). Leide (Leyden) and Düsseldorf 1854.

6 partly hand-coloured, partly colour-printed plates by J. v. Aken, Q. M. R. Ver Huell and A. J. Wendel, lithographed by Arnz & Co.

VRIESE, WILLEM HENDRIK DE

TUINBOUW—FLORA VAN NEDERLAND EN ZIJNE OVERZEESCHE BEZITTINGEN . . .

3 vols 8vo (25 cm × 16 cm). Leiden (Leyden) 1855–56.

35 plates (27 chromolithographed) by J. v. Aken, A. Brouwer, Q.M.R. Ver Huell and A. J. Wendel, lithographed by A. J. Bos, Riocreux, P.W.M. Trap, A. J. Wendel and Wohlfart, together with a number of engraved text-figures.

VRIESE, WILLEM HENDRIK DE

PLANTAE INDIAE BATAVAE ORIENTALIS QUAS, IN ITINERE PER INSULAS ARCHIPELAGI INDICI . . . ANNIS 1815–1821, EXPLORAVIT CASP. GEORG. CAROL. REINWARDT.

2 parts 4to (31 cm × 23½ cm). Lugduni Batavorum (Leyden) 1856 [–1857].

8 plates by A. J. Kouwels, Q.M.R. Ver Huell and A. J. Wendel, lithographed by L. Stroobant. Fasc. 1, pp. 1–80 in 1856, Fasc. 2, pp. 81–160 in 1857.

WALDSTEIN, FRANZ DE PAULA ADAM, GRAF VON, & KITAIBEL, PAUL

DESCRIPTIONES ET ICONES PLANTARUM RARIORUM HUNGARIAE.

3 vols Folio (46 cm × 31½ cm). Viennae [1799–] 1802–12.

280 hand-coloured engraved plates.

For dates of publication see W. T. Stearn in *Flora Malesiana* I. 4: ccxvi (1954).

WALLICH, NATHANIEL (NATHAN WOLFF)

TENTAMEN FLORAE NAPALENSIS ILLUSTRATAE. CONSISTING OF BOTANICAL DESCRIPTIONS AND LITHOGRAPHED FIGURES OF SELECT NIPAL PLANTS.

2 parts Folio (45 cm × 29 cm). Calcutta & Serampore [1824–] 1826.

50 lithographed plates by Gorachaud and Vishnupersaud.

For dates of publication see W. T. Stearn in *Flora Malesiana* I. 4: ccxi (1939).

WALLICH, NATHANIEL (NATHAN WOLFF)

PLANTAE ASIATICAE RARIORES; OR DESCRIPTIONS AND FIGURES OF . . . UNPUBLISHED EAST INDIAN PLANTS.
3 vols Folio (53 cm × 36½ cm). London [1829–] 1830–32.
300 hand-coloured plates (pls 1–295 and map numbered 296–300) by Vishnupersaud, Gorachaud, Miss Drake, W. Griffith and M. Gauci, lithographed by M. Gauci.
For dates of publication see W. T. Stearn in *Flora Malesiana* I, 4: ccxvi (1954).
Vishnupersaud is the most talented of the native Indian artists. W.B.

WANGENHEIM, FRIEDRICH ADAM JULIUS VON

BEYTRAG ZUR TEUTSCHEN HOLZGERECHTEN FORSTWISSENSCHAFT, DIE ANPFLANZUNG NORDAMERICANISCHER HOLZARTEN, MIT ANWENDUNG AUF TEUTSCHE FORSTE BETREFFEND.
Folio (40½ cm × 23½ cm). Göttingen 1787.
31 engraved plates by Wangenheim.

WARNER, ROBERT & WILLIAMS, BENJAMIN SAMUEL

SELECT ORCHIDACEOUS PLANTS.
Folio (45 cm × 32½ cm). London 1862–65.
40 hand-coloured lithographed plates all by J. Andrews and W. H. Fitch.
Ser II. 1865–75.
39 hand-coloured lithographed plates by W. H. Fitch.
Ser III. [1877–91].
39 hand-coloured lithographed plates by W. H. Fitch and J. L. Macfarlane.

WARNER, ROBERT & WILLIAMS, BENJAMIN SAMUEL

THE ORCHID ALBUM, COMPRISING COLOURED FIGURES AND DESCRIPTIONS OF . . . ORCHIDACEOUS PLANTS . . . THE BOTANICAL DESCRIPTIONS BY T. MOORE.
11 vols 4to (30 cm × 24 cm). London 1882–97.
(vols 1–6) by T. Moore. (Vol. 9, conducted by R. Warner, B. S. Williams, H. Williams and W. H. Gower; Vol. 10 conducted by R. Warner, H. Williams and W. H. Gower; Vol. II, by R. Warner and H. Williams).
528 partly hand-coloured, partly colour-printed lithographed plates all by J. N. Fitch.

WATSON, WILLIAM & BEAN, WILLIAM JACKSON

ORCHIDS, THEIR CULTURE AND MANAGEMENT WITH DESCRIPTIONS OF ALL THE KINDS OF GENERAL CULTIVATION.
8vo. London n.d. (1890).
12 chromolithographs. H.B.

WAWRA RITTER VON FERNSEE, HEINRICH

DIE BOTANISCHE ERGEBNISSE DER REISE S. M. DES KAISERS VON MEXICO MAXIMILIAN I NACH BRASILIEN (1859–60).
Folio (48 cm × 33½ cm). Wien (Vienna) 1866.
104 plates (pls 1–32 colour-printed) by J. Sebouth, lithographed by Hartinger u. Sohn.

WAWRA RITTER VON FERNSEE, HEINRICH & BECK VON MANNAGETTA, GÜNTNER

ITINERA PRINCIPUM S. COBURGI. DIE BOTANISCHE AUSBEUTE VON DEN REISEN . . . DER PRINZEN VON SACHSEN-COBURG-GOTHA. I. REISE DER PRINZEN PHILIPP UND AUGUST UM DIE WELT (1872–73). II. REISE DER PRINZEN AUGUST UND FERDINAND NACH BRASILIEN (1879).
2 vols 4to (34½ cm × 27 cm). Wien (Vienna) 1883–88.
57 chromolithographed plates (Vol. 1, pls 1–39, Vol. 2, pls 1–18), by W. Liepoldt, M. v. Schlereth and G. Beck, lithographed by M. Streicher and A. Lenz.
Vol. 1 by Wawra, vol. 2 was completed by G. Beck after Wawra's death, using his notes.

WEBB, PHILIP BARKER-. See BARKER-WEBB, PHILIP

WEDDELL, HUGH ALGERNON

HISTOIRE NATURELLE DES QUINQUINAS, OU MONOGRAPHIE DU GENRE CINCHONA.
Folio (49 cm × 33½ cm). Paris 1849.
33 plates (pls 1–30, frontispiece, 3bis and 4bis) by J. Denis, Riocreux and Steinheil, engraved by Davesne, Daubigny, Mme Gouffé, Picart, Mme Rebel, Steinheil, Mlle E. Tallent and Visto.

WEDDELL, HUGH ALGERNON

CHLORIS ANDINA. ESSAI D'UNE FLORE DE LA RÉGION ALPINE DES CORDILLÈRES DE L'AMÉRIQUE DU SUD . . . (EXPÉDITION . . . PENDANT LES ANNÉES 1843 À 1847, SOUS LA DIRECTION DU COMTE FRANCIS DE CASTELNAU, Part 6, BOTANIQUE).
2 vols 4to (30½ cm × 23 cm). Paris 1855–57 [–1861].
90 plates by A. Riocreux, lithographed by A. Riocreux and Geny-Gros.
For dates of publication see C. D. Sherborn & B. B. Woodward in *Ann. Mag. Nat. Hist.* VII, 8: 164 (1901).

WEIHE, KARL ERNST AUGUST & NEES VON ESENBECK, CHRISTIAN GOTTFRIED

RUBI GERMANICI DESCRIPTI ET FIGURIS ILLUSTRATI. DIE DEUTSCHEN BROMBEERSTRÄUCHE . . .
1 vol. (containing 10 fasc.) Folio (38 cm × 23½ cm). Elberfeld 1822–27.

Pl. 7.
STANHOPEA TIGRINA.

53 partly hand-coloured plates (pls 1–49, 3b, 45b, 46b–c) by W. Engels and T. Wild, engraved by W. Engels, C. Müller, J. Heyer and T. Wild.

WEINMANN, JOHANN WILHELM

PHYTANTHOZA ICONOGRAPHIA . . .

4 vols Folio (39 cm × 24½ cm). Ratisbonae (Regensburg) [1735–] 1737–45.

1025 hand-coloured engraved plates by J. J. Haid, J. E. Ridinger, B. Seuter and G. D. Ehret.

WENDLAND, JOHANN CHRISTOPH

ERICARUM ICONES ET DESCRIPTIONES. ABBILDUNG UND BESCHREIBUNG DER HEIDEN.

3 vols 4to (28½ cm × 23 cm). Hannover 1798–1823.

162 unnumbered hand-coloured engraved plates all by J. C. Wendland.

Vol. 1, fasc. 1–12. 1798–1803. Vol. 2, fasc. 13–24. 1804–10. Vol. 3, fasc. 25–27. 1823.

WENDLAND, JOHANN CHRISTOPH

HORTUS HERRENHUSANUS SUI PLANTAE RARIORUS QUAE IN HORTO REGIO HERRENHUSANO PROPE HANNOVERAM COLUNTUR.

4 parts Folio (46½ cm × 29½ cm). Hannoverae 1798–1801.

24 hand-coloured engraved plates by J. C. Wendland.

A continuation of Schrader, *Sertum Hannoveranum*.

WENDLAND, JOHANN CHRISTOPH

COLLECTIO PLANTARUM TAM EXOTICARUM QUAM INDIGENARUM CUM DELINEATIONE, DESCRIPTIONE CULTURAQUE EARUM. SAMMLUNG AUSLÄNDISCHER U. EINHEIMISCHER PFLANZEN . . .

3 vols 4to (25 cm × 19½ cm). Hannover 1805–19.

84 hand-coloured engraved plates all by J. C. Wendland.

For dates of publication, see O. Kuntze *Revisio Gen. Pl.* 3 (2): 162 (1898).

WENDLAND, JOHANN CHRISTOPH. See SCHRADER, HEINRICH ADOLPH. 1795.

WIGHT, ROBERT

ILLUSTRATIONS OF INDIAN BOTANY, OR FIGURES . . . OF INDIAN PLANTS DESCRIBED IN THE AUTHOR'S PRODROMUS FLORAE PENINSULAE INDIAE ORIENTALIS . . .

2 vols 4to (27½ cm × 21 cm). Madras [1838–] 1840–50.

207 hand-coloured plates by Govindoo, Rungiah, lithographed by Dumphy, Wight, Winchester.

The uppermost number on the plate is its serial number.

WIGHT, ROBERT

ICONES PLANTARUM INDIAE ORIENTALIS, OR FIGURES OF INDIAN PLANTS.

6 vols 4to (27½ cm × 20½ cm). Madras & London (1838–) 1840–53.

2113 plates (pls 1–2101, 35 bis, 122bis, 960^{2}, $964^{2,3}$, 970bis, $995^{2,3}$, 1064bis, 1724bis, 1776bis, 1976bis) by Govindoo, Rungiah, R. Wight and others, lithographed by Dumphy, Romeo, R. Wight and Winchester. For dates of publication, see E. D. Merrill in *J. Arnold Arb*, 22: 222–224 (1941).

WIGHT, ROBERT

SPICILEGIUM NEILGHERRENSE, OR A SELECTION OF NEILGHERRY PLANTS, DRAWN AND COLOURED FROM NATURE, WITH BRIEF DESCRIPTIONS OF EACH . . .

2 vols 4to (27 cm × 21 cm). Madras & Calcutta [1846–] 1851.

203 hand-coloured plates (pls 1–202, 42^{1}, $67^{2,3}$, with pls 74 and 75 missing) by Govindoo and Rungiah, lithographed by Dumphy.

WILHELM, KARL. See HEMPEL, GUSTAV. 1889

WILLDENOW, CARL LUDWIG

HISTORIA AMARANTHORUM.

Folio (41 cm × 25½ cm). Turici (Zurich) 1798.

Frontispiece and 12 hand-coloured engraved plates by J. R. Schellenberg.

WILLDENOW, CARL LUDWIG

PHYTOGRAPHIA, SEU DESCRIPTIO RARIORUM MINUS COGNITARUM PLANTARUM.

Folio (37½ cm × 25 cm). Erlangae (Erlangen) 1794.

10 plates by Stahlknecht, engraved by G. Vogel.

WILLDENOW, CARL LUDWIG

HORTUS BEROLINENSIS SIVE ICONES ET DESCRIPTIONES, PLANTARUM RARIORUM VEL MINUS COGNITARUM, QUAE IN HORTO REGIO BOTANICO BEROLINENSIS EXCOLUNTUR.

2 vols Folio (40½ cm × 25½ cm). Berolini (Berlin) [1803–] 1806–16.

108 hand-coloured plates by F. Guimpel, engraved by

PLATE 47

James Bateman's Orchidaceae of Mexico and Guatemala *(London 1837–43) uses huge hand-colored lithographs measuring 31″ × 20½″ for its fifty plates. Mrs. Withers drew this dramatic orchid, and M. Gauci transcribed it onto stone. This weighty tome confirmed that orchidomania had indeed taken root in England.*

F. Guimpel, P. Haas and L. Schmidt.
For dates of publication, see W. T. Stearn in *J. Bot. (London)* 75: 233–35 (1937).

WILLDENOW, CARL LUDWIG & HAYNE, FRIEDRICH GOTTLOB

ABBILDUNG DER DEUTSCHEN HOLZARTEN FÜR FORSTMÄNNER UND LIEBHABER DER BOTANIK.
2 vols 4to (26 cm × 20½ cm). Berlin 1815–20.
216 hand-coloured engraved plates all by F. Guimpel.
After the death of Willdenow in 1812, Hayne undertook the completion of the work under the editorship of Guimpel.

WILLKOMM, HEINRICH MORITZ

ICONES ET DESCRIPTIONES PLANTARUM NOVARUM CRITICARUM ET RARIORUM EUROPAE AUSTRO-OCCIDENTALIS PRAECIPUE HISPANIAE.
2 vols 4to (33 cm × 25½ cm). Lipsiae (Leipzig) 1852–56 [–62].
168 hand-coloured plates by H. M. Willkomm, engraved by A. H. Payne.
For dates of publication, see F. G. Wiltshear in *J. Bot. (London)* 53: 370 (1915).

WILLKOMM, HEINRICH MORITZ

ILLUSTRATIONES FLORAE HISPANIAE INSULARUMQUE BALEARIUM. FIGURAS DE PLANTAS NUEVAS O RARAS . . . FIGURES DE PLANTES NOUVELLES OU RARES . . .
2 vols Folio (34 cm × 25 cm). Stuttgart 1881–92.
183 hand-coloured plates by H. M. Willkomm, lithographed by A. Eckstein and G. Ebenhusen.
For dates of publication, see F. G. Wiltshear in *J. Bot (London)* 53: 372 (1915).

WOODVILLE, WILLIAM

MEDICAL BOTANY WITH PLATES OF ALL THE MEDICINAL PLANTS . . . IN THE CATALOGUES OF THE MATERIA MEDICA . . .
3 vols and supplement. Folio (22 cm × 17½ cm). London 1790–93. Supplement, 1794–(95).
274 hand-coloured engraved plates (pls 1–210; Suppl., 211–274) all by J. Sowerby.
Ed. 2. 4 vols 4to (26½ cm × 19 cm). London (1805) 1810.
274 hand-coloured engraved plates by J. Sowerby.
Ed. 3 . . . the botanical descriptions arranged and corrected by W. J. Hooker . . . the new medico-botanical portion supplied by G. Spratt . . . 5 vols 4to (25½ × 18½ cm). London 1832.
274 hand-coloured engraved plates as in ed. 2 (forming vols 1–4) by J. Sowerby, and 36 hand-coloured engraved plates undated (forming Vol. 5) by G. Spratt. According to the publisher's advertisement this edition was limited to 200 copies.
In Vol. 2 of eds. 2 and 3, plate 112 is missing.
In Vol. 3 of ed. 2, there are two plates numbered 181, the *second* representing *Tormentilla erecta*.
In Vol. 3 of ed. 3, there are two plates numbered 182, the *first* representing *Tormentilla erecta*.

WOOLWARD, FLORENCE H.

THE GENUS MASDEVALLIA. ISSUED BY THE MARQUIS OF LOTHIAN, CHIEFLY FROM PLANTS . . . AT NEWBATTLE ABBEY; PLATES AND DESCRIPTIONS BY F. H. WOOLWARD; WITH ADDITIONAL NOTES BY F. C. LEHMANN.
Folio (43½ cm × 31 cm). Grantham [1890–] 1896.
87 hand-coloured lithographed unnumbered plates by F. H. Woolward and F. C. Lehmann (1 only) and 61 woodcuts from photographs.

WOOSTER, DAVID

ALPINE PLANTS: FIGURES AND DESCRIPTIONS OF SOME OF THE MOST STRIKING AND BEAUTIFUL OF THE ALPINE FLOWERS.
2 vols 8vo (25 cm × 17½ cm). London (1871–) 1872–74.
108 hand-coloured engraved plates (1–54 1–54), by A. F. Lydon.
Ed. 2 (1st vol. only) 1874.
2nd vol. of 1st ed. is called '2nd Series'. 1871 in preface, 1872, 1874 on title pages.

[YONGE, CHARLOTTE M.]

THE INSTRUCTIVE PICTURE BOOK OF LESSONS FROM THE VEGETABLE WORLD.
BY THE AUTHORESS OF THE HEIR OF REDCLIFFE.
Edinburgh 1857.
31 double pages of hand-coloured lithographs.
Published in pictorial boards. A number of editions.
H.B.

ZENKER, JONATHAN CARL

PLANTAE INDICAE, QUAS IN MONTIBUS COIMBATURICIS COERULEIS, NILAGIRI SEU NEILGHERRIES DICTIS, COLLEGIT REV. B. SCHMIDT.
2 parts Folio (40 cm × 25 cm). Jena and Paris 1835–37.
20 hand-coloured plates by J. C. Zenker, engraved by F. Kirchner and W. Müller.

[ZORN, JOHANNES]

ICONES PLANTARUM MEDICINALIUM. ABBILDUNGEN VON ARZNEYGEWÄCHSEN.
6 vols 8vo (19½ cm × 12 cm). Nürnberg (Nuremberg) 1779–90.
600 hand-coloured engraved plates by B. Thanner (and others?), engraved by J. M. Burucker, J. C. Claussner, J. S. Leitner, J. K. Mayr, J. C. Pemsel.
Vol. 1, pls 1–100, was reissued in 1784 as 'Zweyte Auflage'.

Pl. 45.

Miss Drake del.

M. Gauci, lith.

Epidendrum vitellinum.

Pub.d by J. Ridgway & Sons 169 Piccadilly Dec.r 1841.

Printed by P. Gauci

ANDREWS, HENRY C.

THE BOTANIST'S REPOSITORY, FOR NEW, AND RARE PLANTS. CONTAINING COLOURED FIGURES OF SUCH PLANTS, AS HAVE NOT HITHERTO APPEARED IN ANY SIMILAR PUBLICATION; WITH ALL THEIR ESSENTIAL CHARACTERS, BOTANICALLY ARRANGED, AFTER THE SEXUAL SYSTEM OF THE CELEBRATED LINNAEUS; IN ENGLISH, AND LATIN. TO EACH DESCRIPTION IS ADDED, A SHORT HISTORY OF THE PLANT, AS TO ITS TIME OF FLOWERING, CULTURE, NATIVE PLACE OF GROWTH, WHEN INTRODUCED, AND BY WHOM. THE WHOLE EXECUTED BY HENRY ANDREWS . . .
(An additional engraved title-page in each volume, as follows: THE BOTANISTS REPOSITORY, COMPRISING COLOUR'D ENGRAVINGS OF NEW AND RARE PLANTS ONLY WITH BOTANICAL DESCRIPTIONS ETC. IN LATIN AND ENGLISH AFTER THE LINNAEAN SYSTEM, BY H. ANDREWS).
10 vols 4to (27 cm × 20½ cm). T. Bensley, London 1797–1814.
664 hand-coloured engraved plates, all by H. C. Andrews.
For dates of publication, see J. Britten in *J. Bot. (London)* 54: 236–246 (1916), *Flora Malesiana* I. 4: clxvi. The original wrappers have different wording to either title-pages; this reads as follows:
The Botanist's Repository for new and rare plants only. A work designed to comprise coloured figures of each plant, as have not hitherto been given to the public in any similar publication.
According to Britten, the text of the first 5 volumes was probably written by John Kennedy but vol. 6 was written by A. H. Haworth; see Stearn in *Gard. Chron.* III. 116: 40 (1944). George Jackson supervised later volumes.

ANNALES DE LA SOCIÉTÉ ROYALE D'AGRICULTURE ET DE BOTANIQUE DE GAND. See GHENT.

BOTANIC GARDEN. See MAUND, BENJAMIN.

BOTANICAL CABINET. See LODDIGES, CONRAD, & SONS.

BOTANICAL MAGAZINE. See CURTIS, WILLIAM, & OTHERS.

BOTANICAL REGISTER. See EDWARDS, SYDENHAM, & OTHERS.

BOTANIST, THE. See MAUND, BENJAMIN, & HENSLOW, JOHN STEVENS.

BOTANIST'S REPOSITORY. See ANDREWS, HENRY.

BRITISH FLORIST, THE

6 vols Post 8vo. London 1846.
81 hand-coloured lithographs. H.B.

BROOKSHAW, GEORGE

THE HORTICULTURAL REPOSITORY,
containing delineations of the best varieties of the different species of English fruits: to which are also added, delineations of the blossoms and leaves, in those instances in which they are considered necessary; together with descriptions of each fruit . . .
2 vols 8vo (24 cm × 15½ cm). London (1820–1823).
104 hand-coloured engraved plates by G. Brookshaw.
Some large folding plates are given two consecutive plate-numbers.

CALCUTTA, ROYAL BOTANIC GARDEN

ANNALS OF THE ROYAL BOTANIC GARDEN.

Vol. 1. KING, GEORGE. THE SPECIES OF FICUS OF THE INDO-MALAYAN AND CHINESE COUNTRIES.
Folio (37 cm × 27 cm). Calcutta & London 1887–88.
238 uncoloured plates in Parts I and II (pls 1–232, 2a, 2b, 30a, 43a, 58a, 95a) by G. C. Das., M. Smith, A. L. Singh, A. C. Singh, and S. Ab Dool Mullah, lithographed by many artists.
Appendix (included in Part II):
SOME NEW SPECIES OF FICUS FROM NEW GUINEA.
1 photo-etching and 4 uncoloured lithographed pls by D. D. Cunningham.

Vol. 2. KING, GEORGE. THE SPECIES OF ARTOCARPUS INDIGENOUS TO BRITISH INDIA; THE INDO-MALAYAN SPECIES OF QUERCUS AND CASTANOPSIS.
Folio (34½ cm × 27 cm). Calcutta 1889.
Plates by G. C. Das, A. L. Singh, M. Smith and A. D. Mulla, lithographed by many artists.

Vol. 3. PRAIN, DAVID. THE SPECIES OF PEDICULARIS OF THE INDIAN EMPIRE AND ITS FRONTIERS.
(Plates 1–37).
KING, GEORGE. THE MAGNOLIACEAE OF BRITISH INDIA.
(Plates 38–74, 47 bis, 47 ter, 57 bis).

PLATE 48

John Lindley's Sertum orchidaceum *(London 1837–41) illustrated new orchids discovered in South America. "Epidendrum vitellinum," pictured here, had already been illustrated in Curtis's* Botanical magazine; *Lindley justified featuring his own robust plant by denigrating the Curtis print as a flawed rendering of a weak specimen. Competition to publish detailed drawings and name the new beauties (at times, after oneself) was integral to orchidomania.*

PRAIN, DAVID. AN ACCOUNT OF THE GENUS GOMPHOSTEMMA WALL.

(Plates 75–105).

KING, GEORGE. THE SPECIES OF MYRISTICA OF BRITISH INDIA.

(Plates 106–174, 107bis, 124bis, 124ter, 125bis, 141bis).

Folio (34½ cm × 27 cm) Calcutta (1890–) 1891.

182 uncoloured plates by A. L. Singh, A. D. Mulla (Molla), G. C. Das, K. P. Das, D. Prain, M. Smith, D. N. Chaudhary, A. D. Chaudhary, W. D. Alwis, lithographed by many artists. Dated 1890.

Vol. 4. KING, GEORGE. THE ANONACEAE OF BRITISH INDIA.

Folio (54½ cm × 27 cm). Calcutta & London 1893.

225 uncoloured plates (pls 1–220, 20bis, 116bis, 116ter, 169bis, 208bis) by K. P. Das, G. C. Das, A. D. Mulla (Molla), D. N. Chaudhary, A. L. Singh, M. Smith, R. K. Das, W. D. Alwiss, T. Alwiss, G. de Alwis, H. Alwiss, J. D. Alwiss, A. D. Alwiss, Govindeo, lithographed by many artists.

Vol. 5. Part I HOOKER, J. D. A CENTURY OF INDIAN ORCHIDS.

(Plates 1–101).

Part II. BRÜHL, PAUL JOHANNES & KING, GEORGE. A CENTURY OF NEW AND RARE INDIAN PLANTS. (PLATES 102–200, 110A). TO WHICH IS PREFIXED A BRIEF MEMOIR OF WILLIAM ROXBURGH BY G. KING.

Folio (34½ cm × 27 cm) Calcutta & London 1895–96.

201 plates (of which pls 1–101 are hand-coloured) by A. L. Singh, G. C. Das, J. D. Hooker, M. Smith, G. King, G. E. Hooker, K. P. Das, D. A. Chaudhary, A. D. Mulla (Molla), R. Pantling and D. Prain, lithographed by many artists.

Vol. 6. Part I. CUNNINGHAM, D. D. 1. THE CAUSES OF FLUCTUATIONS IN TURGESCENCE IN THE MOTOR-ORGANS OF LEAVES. (PLATES 1–7). 2. A NEW AND PARASITIC SPECIES OF CHOANEPHORA. (Plates 8, 9).

Folio (34½ cm × 27 cm). Calcutta & London 1895.

9 hand-coloured plates by D. D. Cunningham, lithographed by A. C. Mukerjee, P. N. Sinha, A. C. Sinha, K. D. Chandra and S. C. Mondul.

Part II. WEST, WILLIAM & WEST, GEORGE STEPHEN. FRESH-WATER ALGAE FROM BURMA, INCLUDING A FEW FROM BENGAL AND MADRAS. (PLATES 10–16).

Folio (34½ cm × 27 cm). Calcutta 1907.

7 plates by C. S. West, lithographed by Hull.

Vol. 7. GAMBLE, JAMES SYKES. THE BAMBUSEAE OF BRITISH INDIA.

Folio (34½ cm × 27 cm). Calcutta & London 1896.

119 uncoloured plates by M. Idrees, K. P. Das, G. C. Das, A. L. Singh, A. D. Molla, Hormasjie, D. N. Choudhury, lithographed by many artists.

Vol. 8. KING, GEORGE & PANTLING, ROBERT. THE ORCHIDS OF THE SIKKIM-HIMALAYA. Part I, text; Parts II–IV, plates.

Folio (34½ cm × 27½ cm). Calcutta and London 1898.

453 hand-coloured plates (pls 1–448, 244A, 402bis, 404bis, 430bis, 444bis) by R. Pantling, lithographed by many artists.

Volumes 9 and onwards of this work, which continued till 1942, were published after 1900.

CURTIS, WILLIAM, & OTHERS.

THE BOTANICAL MAGAZINE, OR FLOWER-GARDEN DISPLAYED: IN WHICH THE MOST ORNAMENTAL FOREIGN PLANTS, CULTIVATED IN THE OPEN GROUND, THE GREENHOUSE, AND THE STOVE, ARE ACCURATELY REPRESENTED IN THEIR NATURAL COLOURS. TO WHICH ARE ADDED THEIR NAMES . . . TOGETHER WITH THE MOST APPROVED METHODS OF CULTURE. BY WILLIAM CURTIS.

Vols 1–14 8vo (23 cm × 14 cm). London (1787–) 1790–1800.

CURTIS'S BOTANICAL MAGAZINE . . .

CONTINUED BY JOHN SIMS.

Vols 15–42 8vo (23 cm × 14 cm). London 1801–1815.

Vols 43–53 (New Series I, vols 1–11).

8vo (23 cm × 14 cm). London 1816–26.

2704 hand-coloured plates by John Curtis, S. T. Edwards, R. K. Greville, W. J. Hooker, W. Herbert, W. Kilburn and J. Sowerby, engraved by W. Darton, F. Sansome, J. Sowerby, Swan and Weddell.

Vols 54–70 (New Series II, vols 1–17) . . . conducted by Samuel Curtis, the descriptions by William Jackson Hooker.

8vo (23½ cm × 14½ cm). London 1827–44.

1426 hand-coloured plates (pls 2705–4131) by M. Curtis, W. Curtis, W. Fitch, Greville, Harrison, W. J. Hooker and Mrs Withers, engraved by Swan.

Vols 71–90, (Third Series, vols 1–20) . . . comprising the plants of the Royal Gardens of Kew, and of other botanical establishments in Great Britain with suitable descriptions . . . by Sir William Jackson Hooker.

Vols 91–130. (Third Series, vols 21–60) . . . with suitable descriptions by Joseph Dalton Hooker, assisted (in vols 129, 130) by William Botting Hemsley.

8vo (24 cm × 14½ cm). London 1845–1904.

3859 hand-coloured plates (pls 4132–7991) by H. Thiselton-Dyer, W. Fitch and M. Fitch, engraved by

Swan or lithographed by W. Fitch and J. N. Fitch. A complete index, with modern names for the plants figured, published by The Royal Horticultural Society in 1956.

In 1948, when the Botanical Magazine had completed its 164th volume with the publication of pl. 9688, a new Series with plates numbered from N.S. 1 onwards began.

EDWARDS, SYDENHAM & OTHERS

THE BOTANICAL REGISTER: CONSISTING OF COLOURED FIGURES OF EXOTIC PLANTS, CULTIVATED IN BRITISH GARDENS; WITH THEIR HISTORY AND MODES OF TREATMENT. THE DESIGNS BY SYDENHAM EDWARDS [from vol. 5 'and others'].
Vols 1–13 1815–27.

THE BOTANICAL REGISTER; OR ORNAMENTAL FLOWER-GARDEN AND SHRUBBERY. CONSISTING OF COLOURED FIGURES OF PLANTS AND SHRUBS CULTIVATED IN BRITISH GARDENS; ACCOMPANIED BY THEIR HISTORY, BEST METHODS OF TREATMENT IN CULTIVATION, PROPAGATION, ETC.
Vol. 14, New Series, Vol. 1. 1828.

EDWARD'S BOTANICAL REGISTER . . . CONTINUED BY J.L. LINDLEY.
Vols 15–33, New Series, Vols 2–20. 1829–47.
8vo cm (23 × 14½ cm). London 1815–47.
2710 hand-coloured engraved plates, and 1 unnumbered, uncoloured, principally by Sydenham Edwards, M. Hart and Miss Drake; others by W. B. Booth, J. T. Hart, W. Herbert, J. Lindley, C. J. Robertson, E. D. Smith, and 15 other artists who contributed 1 or 2 plates each. Volumes 1–23 contain 2022 plates (pls. 1–2014, with two plates each to numbers 600, 700, 706, 720, 736, 750, 774, and 1203); from Volume 24, New Series 11, onwards, each volume has plates numbered separately (Vol. 24 pls 1–68; Vol. 25 pls 1–69; Vol. 26 pls 1–71; Vol. 27 pls 1–70; Vol. 28 pls 1–69; Vol. 29, pls 1–66; Vol. 30 pls 1–67; Vol. 31 pls 1–69, Vol. 32. pls 1–69, Vol. 33 pls 1–70).

APPENDIX TO THE FIRST TWENTY-THREE VOLUMES OF EDWARDS'S BOTANICAL REGISTER: CONSISTING OF A COMPLETE INDEX . . . WITH A SKETCH OF THE VEGETATION OF THE SWAN RIVER COLONY . . . BY J. LINDLEY . . .
London 1839 [–40].
9 hand-coloured lithographed plates and 4 uncoloured engraved text figures.
According to W. T. Stearn in *J. Soc. Bibl. Nat. Hist.* 2: 381 (1952), issued in 3 parts: part 1, Appendix i–xvi, pls 1–4, Index i–xxxii, 1 Nov. 1839; part 2, Appendix xvii–xxi; pls 5–7, Index xxxiii–xlviii; part 3, Appendix xxxiii–lviii, pls 8–9, Index xlix–lxiv, 1 Jan. 1840.

ENCYCLOGRAPHIE DU RÈGNE VÉGÉTAL

PRÉSENTANT LA FIGURE, LA DESCRIPTION ET L'HISTOIRE DES PLANTES LE PLUS RÉCEMMENT DÉCOUVERTES SUR TOUS LES POINTS DU GLOBE OU INTRODUITES DANS LES SERRES DES JARDINS D'ANGLETERRE, DE LA BELGIQUE ET DES AUTRES PARTIES DE L'EUROPE, ACCOMPAGNÉE DE MONOGRAPHIES DE GENRES DESTINÉES À FORMER PROGRESSIVEMENT UNE FLORE UNIVERSELLE. OUVRAGE PUBLIÉ SOUS LA DIRECTION DE M. DRAPIEZ (TOME CINQUIÈME 'SOUS LA DIRECTION D'UNE SOCIETÉ DE BOTANISTES').
6 vols bound in 4. Folio (37 cm × 28½ cm). Bruxelles (Brussels) 1833–1837 (1838).
370 unnumbered hand-coloured engraved and lithographed plates copied from Edwards, Andrews, Redouté and others.

FLORAL MAGAZINE, THE

COMPRISING FIGURES AND DESCRIPTIONS OF POPULAR GARDEN FLOWERS. BY T. MOORE (Vol. 1), THEREAFTER BY H. H. DOMBRAIN.
10 vols 8vo (24½ cm × 16½ cm). London 1861–71.
560 hand-coloured lithographed plates by W. H. Fitch, J. Andrews and J. Worthington Smith.

NEW SERIES . . . FIGURES AND DESCRIPTIONS OF THE CHOICEST NEW FLOWERS FOR THE GARDEN, STOVE, OR CONSERVATORY, BY H. H. DOMBRAIN (Vols 1–4), F. W. BURBIDGE (Vol. 5), RICHARD DEAN (Vols 6–10).
10 vols 4to (30 cm × 23½ cm). London 1872–81.
480 hand-coloured lithographed plates by W. G. Smith, F. W. Burbidge and J. N. Fitch.

FLORE DE L'AMATEUR

CHOIX DES PLANTES PUBLIÉES DANS LE SERTUM BOTANICUM (Q.V.)
Folio. Brussels n.d. (1833).
200 hand-coloured lithographs. H.B.

FLORE DE L'AMATEUR

CHOIX DES PLANTES LES PLUS REMARQUABLES PAR LEUR ÉLÉGANCE ET LEUR UTILITÉ . . . CONTENANT 170 PLANCHES . . . AVEC TEXTE EXPLICATIF.
2 vols Folio (34½ cm × 25½ cm). Paris 1847.
177 unsigned lithographed plates, colour-printed and re-touched by hand.
This is a re-issue of selected plates with text from the *'Flore des Serres . . . de Paris'*.

FLORE DES SERRES ET DES JARDINS DE L'EUROPE

OU DESCRIPTIONS DES PLANTES LES PLUS RARES ET LES PLUS MÉRITANTES, NOUVELLEMENT INTRODUITES SUR LE CONTINENT OU EN ANGLETERRE, OUVRAGE ENRICHI DE NOTICES

J. D. H. del. Fitch, lith.

RHODODENDRON FULGENS, Hook. fil.

HISTORIQUES, HORTICULTURALES, ETC., ET RÉDIGÉE PAR CH. LEMAIRE, M. SCHEIDWEILER ET L. VAN HOUTTE.
Vols 1–10 8vo (25 cm × 17 cm). Gand (Ghent) 1845–55.
(143 unnumbered plates, and pls. 143–1076).

. . . JOURNAL GÉNÉRAL D'HORTICULTURE, SOUS LA DIRECTION DE J. DECAISNE ET L. VAN HOUTTE (Vols 11–15).
Vols 11–15 (of which 11–14 are called 2nd Série, 1–4) 8vo (25 cm × 17 cm). Gand (Ghent). 1856–65. (Plates 1077–1607).
2471 coloured and 56 uncoloured lithographed plates and many engraved text-figures.

ANNALES GÉNÉRALES D'HORTICULTURE.
Vols 16–23 Gand (Ghent) 1865–83 (Plates 1608–2471).

FLORE DES SERRES ET JARDINS DE PARIS
OU COLLECTION DE PLANTES, REMARQUABLES PAR LEUR UTILITÉ, LEUR ÉLÉGANCE, LEUR ÉCLAT OU LEUR NOUVEAUTÉ . . . SIX CENT PLANCHES ACCOMPAGNÉES D'UN TEXTE PARTICULIER POUR CHAQUE PLANTE . . . PAR UNE SOCIÉTÉ DE BOTANISTES.
6 vols Folio (37 cm × 27 cm). Paris 1834.
600 unsigned hand-coloured lithographed plates.

FLORIST, THE
Vol. 1 8vo (22 cm × 14 cm). London 1848.
continued as THE FLORIST AND GARDEN MISCELLANY.
2 vols 8vo (22 cm × 14 cm). London 1849–50.
continued as THE FLORIST, FRUITIST AND GARDEN MISCELLANY.
11 vols 8vo (22 cm × 14 cm). London 1851–61.
continued as THE FLORIST AND POMOLOGIST.
23 vols 8vo. London 1862–1877 (24 cm × 15 cm), 1878–84 (27 cm × 18 cm).

GHENT, SOCIÉTÉ ROYALE D'AGRICULTURE ET BOTANIQUE
ANNALES DE LA SOCIÉTÉ ROYALE D'AGRICULTURE ET DE BOTANIQUE DE GAND, JOURNAL D'HORTICULTURE ET DES SCIENCES ACCESSOIRES, RÉDIGÉ PAR CHARLES MORREN.
5 vols 4to (25 cm × 16½ cm). Gand (Ghent), Bruxelles (Brussels) Liège 1845–49.
291 hand-coloured plates by A. Lagarde, C. Morren, M. Morren, L. Berehmans, B. Léon, G. Severeyns, lithographed by G. Severeyns, A. Lagarde, H.v. Geert.

HERBIER DE L'AMATEUR DE FLEURS
CONTENANT, GRAVÉS ET COLORIÉS D'APRÈS NATURE, LES VÉGÉTAUX, QUI PEUVENT ORNER LES JARDINS ET LES SERRES; . . . AVEC UN PRÉCIS D'ORGANOGRAPHIE ET DE PHYSIQUE VÉGÉTALES . . . PAR DRAPIEZ.
8 vols 4to (26½ cm × 20½ cm). Bruxelles (Brussels) 1828–35.
600 hand-coloured plates mostly by Bessa, engraved by Lejeune, Barrois, Goulet, Dennel, Coignet, Guyard fils and others.

HERBIER GÉNÉRAL DE L'AMATEUR
CONTENANT LA DESCRIPTION, L'HISTOIRE, LES PROPRIÉTÉS ET LA CULTURE DES VÉGÉTAUX UTILES ET AGRÉABLES; DÉDIÉ AU ROI, PAR MORDANT DELAUNAY; CONTINUÉ PAR LOISELEUR DESLONGCHAMPS, AVEC FIGURES PEINTES D'APRÈS NATURE.
8 vols 4to (27 cm × 17 cm). Paris [1810–] 1816–27.
575 hand-coloured plates (pls 1–572, with 2 each numbered 171, 172 and 199) by Bessa, Redouté and Poiteau, engraved by 21 engravers.

HERINCQ, F. See L'HORTICULTEUR FRANÇAIS

L'HORTICULTEUR FRANÇAIS DE MIL CENT CINQUANTE ET UN
JOURNAL DES AMATEURS ET DES INTÉRÊTS HORTICOLES. RÉDIGÉ PAR F. HERINCQ.
21 vols 8vo (23½ cm × 15 cm). Paris 1851–72.
414 hand-coloured plates, mainly by E. Bricogne, lithographed by Visto from 1851–59, and by Maubert and Faguet, lithographed by Debray from 1860–72. Other artists: Vianne, J. Migneaux, C. Nivelet, Christiaenssens, Chabal-Dussurgey, Oudart, J. C. Werner, Anceau, 'F.H.', A. Nori, E. Blanchard, H. Laluye, J. Rupalley, L. Clemens, E. Grabowski and Courtin. The work contains also some uncoloured engraved text-figures and full-plate diagrams.
There are no volume numbers except for the years 1859–61, which volumes are called 'Deuxième Série, 1–3'. It is then dropped from the title-page, and continued on the *plates* until 1865, when any reference to volumes or series is discontinued.

PLATE 49
Joseph Hooker botanized in Asia, discovering many hardy shrubs. "Rhododendron Fulgens" from his Rhododendrons of Sikkim-Himalaya *(London 1849–51) was drawn on stone by Walter Fitch while the explorer remained afield. Fitch published thousands of expert drawings of exotics, yet he never left England. Note the artifice of the simultaneous appearance of unopened buds and mature blooms on the same branch.*

HORTICULTURAL JOURNAL AND FLORISTS' REGISTER, THE

. . . (from Vol. 3 'AND ROYAL LADY'S MAGAZINE'). [edited by G. Glenny].
8 vols 8vo (21½ cm × 13½ cm). London. 1834–1837.
c. 63 hand-coloured engraved plates and 3 uncoloured mainly by J. Wakeling, engraved by E. S. Weddell. Uncoloured engraved portraits and plans.
New Series edited by G. Glenny, 2 vols 8vo (24½ cm × 14½ cm). London 1838–40.
29 hand-coloured engraved plates, 1 uncoloured, mainly by J. Wakeling, engraved by E. S. Weddell.

HORTICULTURAL SOCIETY OF LONDON

TRANSACTIONS.
7 vols 4to (27½ cm × 22 cm). London [1807–] 1812–30 [–31].
2nd series. 3 vols 4to. London 1831–38.
In all 175 mostly hand-coloured plates (FIRST SERIES, Vol. 1, pls 1–16; Vol. 2, pls 1–30, 6 (Verd. Grape), 28 (Dimacarpus, Alex Apple); Vol. 3, pls 1–15; Vol. 4; pls 1–22; Vol. 5, pls 1–20, 17; Vol. 6, pls 1–10; Vol. 7, pls 1–15; SECOND SERIES, Vol. 1, pls 1–18; Vol. 2, pls 1–16; Vol. 3, pls 1–5 by Franz Bauer, Lady Boughton, W. Clarke, B. Cotton, Miss Drake, W. Hooker, J. Lindley, C. J. Robertson, E. D. Smith and Mrs Withers, engraved by G. Barclay, E. Clark, W. Clark, C. Fox, W. Hooker, W. Say, B. Taylor and Weddell.
For dates of publication, see W. T. Stearn in *Flora Malesiana* I. 4: ccxiv (1954). c.f. P. M. Synge. 'The publications of the Royal Horticultural Society, 1804–1854' in *J. Roy. Hort. Soc.* 79: 528–536. 1954.

JARDIN FLEURISTE, LE

JOURNAL GÉNÉRAL DES PROGRÈS ET DES INTÉRÊTS HORTICOLES ET BOTANIQUES, CONTENANT L'HISTOIRE, LA DESCRIPTION ET LA CULTURE DES PLANTES LES PLUS RARES ET LES PLUS MÉRITANTES NOUVELLEMENT INTRODUITES EN EUROPE, RÉDIGÉ PAR CH. LEMAIRE.
4 vols 8vo (24½ cm × 16 cm). Gand (Ghent) [1850–] 1851–54.
430 unsigned hand-coloured lithographed plates, and many uncoloured engraved text-figures.
Some are folding plates comprising two ordinary pages and given two plate numbers.

LEMAIRE, CHARLES. See LE JARDIN FLEURISTE.

LINDENIA

ICONOGRAPHIE DES ORCHIDÉES. DIRECTEUR, J. LINDEN. RÉDACTEURS EN CHEF, LUCIEN LINDEN ET EMILE RODIGAS.
Vols 1–10 (Plates 1–480). Folio (35½ cm × 27 cm). Gand (Ghent) 1885–1894.
Vols 11–17, Second Série 1–7 (plates 481–814, with pl. 759 not published), dirigée par J. Linden et Lucien Linden (Vols 12 & 13, Seconde Série 2, 3); dirigée par Lucien Linden (from Vol. 14, Seconde Série, 4). Folio (35½ cm × 27 cm). Gand (Ghent). 1895–1901.
814 chromolithographed plates by P. de Pannemaeker, P. de Tohenaere, E. Bungeroth, C. Triest, A. Goossens, J. Macfarlane, G. Putzys, Henri Leroux, J. de Bosschere, Jeanne Mercier, Alice Mercier, Georgette Meunier, E. Pannech and C. de Bruyne, lithographed by P. de Pannemaeker et fils, G. Severeyns, J. Goffart and S. de Leeuw. Engraved portrait of J. Linden in vol. 13.

LODDIGES, CONRAD, & SONS

THE BOTANICAL CABINET CONSISTING OF COLOURED DELINEATIONS OF PLANTS FROM ALL COUNTRIES, WITH A SHORT ACCOUNT OF EACH, DIRECTIONS FOR MANAGEMENT ETC.
20 vols 4to (21 cm × 16 cm). London 1817–33.
2000 hand-coloured plates (and 1 uncoloured No. 684*) by G. Loddiges, W. Loddiges, Miss J. Loddiges, G. Cooke, E. W. Cooke, W. I. Cooke, T. Boys, W. Miller and Miss Rebello, engraved by G. Cooke.
Vol. 11, pls 1101–1110, July 1826; pls 1111–1120, Aug. 1826; pls 1121–30, Sept. 1826; pls 1131–1140, Oct. 1826; pls 1141–1150, Nov. 1826; pls 1151–1160, Dec. 1826; pls 1161–1170, Jan. 1827; pls 1171–1180, Feb. 1827; pls 1181–1190, March 1827; pls 1191–1200, April 1827.

MAGAZIN DER AESTHETISCHEN BOTANIK. See REICHENBACH, HEINRICH GOTTLIEB LUDWIG.

MAUND, BENJAMIN

THE BOTANIC GARDEN, CONSISTING OF HIGHLY FINISHED REPRESENTATIONS OF HARDY ORNAMENTAL FLOWERING PLANTS CULTIVATED IN GREAT BRITAIN.
13 vols 4to. London 1825–51.
312 hand-coloured plates, each with 4 figures, by Mrs E. Bury, Miss S. Maund, Miss E. Maund, E. D. Smith and Mills, engraved by S. Watts.

MAUND, BENJAMIN, & HENSLOW, JOHN STEVENS

THE BOTANIST; CONTAINING ACCURATELY COLOURED FIGURES, OF TENDER AND HARDY ORNAMENTAL PLANTS, WITH DESCRIPTIONS, SCIENTIFIC AND POPULAR . . . CONDUCTED BY B. MAUND, ASSISTED BY THE REV. J. S. HENSLOW.
5 vols 4to (23 cm × 18½ cm). London [1836–] 1842.
250 hand-coloured unnumbered plates by Miss Taylor, Miss S. Maund, Miss Mintern, Miss Nicholson, Miss Hall, Mills, J. S. Henslow, Bourne, L. Pope, S. Humble, Mrs E. Bury, Mrs Withers, engraved by S. Watts, Nevitt and Smith.

PAXTON, SIR JOSEPH

MAGAZINE OF BOTANY AND REGISTER OF FLOWERING PLANTS.

16 vols 4to. London 1834–49.

768 coloured plates drawn and engraved by F. W. Smith or drawn and lithographed by S. Holden.

Issued in monthly parts; only the last part appeared in the year stated on the title page.

For dates of publication of vols 12 and 13, see W. T. Stearn in *J. Soc. Bibl. Nat. Hist.* 3: 103 (1955). Vol. 1 was reissued in 1841.

REFUGIUM BOTANICUM. See SAUNDERS, WILLIAM WILSON.

SAUNDERS, WILLIAM WILSON

REFUGIUM BOTANICUM; OR, FIGURES AND DESCRIPTIONS FROM LIVING SPECIMENS OF LITTLE KNOWN OR NEW PLANTS OF BOTANICAL INTEREST. EDITED BY W. WILSON SAUNDERS, THE DESCRIPTIONS BY H. G. REICHENBACH, J. G. BAKER AND OTHER BOTANISTS.

5 vols 8vo (24½ cm × 15 cm). London 1869–73.

360 partially hand-coloured lithographed plates all by W. H. Fitch.

SERTUM BOTANICUM

COLLECTION CHOISIE DE PLANTES REMARQUABLES PAR LEUR UTILITÉ, LEUR ÉLÉGANCE, LEUR ÉCLAT OU LEUR NOUVEAUTÉ; CONSISTANT EN SIX CENTS PLANCHES . . . ACCOMPAGNÉES D'UN TEXTE PARTICULIER POUR CHAQUE PLANTE (ENCYCLOGRAPHIE DU RÈGNE VÉGÉTAL) . . . PAR UNE SOCIÉTÉ DE BOTANISTES ET DIRIGÉE PAR P. C. VAN GÉEL.

4 vols. Folio (36 cm × 26½ cm). Bruxelles (Brussels) 1828–32.

600 unnumbered hand-coloured and 1 uncoloured lithographed plates by G. Severeyns, lithographed by de Burggraaff.

VERSCHAFFELT, ALEXANDRE, & VERSCHAFFELT, AMBROISE

NOUVELLE ICONOGRAPHIE DES CAMELLIAS, CONTENANT LES FIGURES ET UNE COURTE DESCRIPTION DES PLUS RARES, DES PLUS NOUVELLES, ET DES PLUS BELLES VARIÉTIÉS DE CE GENRE (ILLUSTRÉE ET RÉDIGÉE PAR UNE SOCIÉTÉ D'AMATEURS ET DE PRACTICIENS—Vol. 1 only).

13 vols 4to (25 cm × 17 cm). Gand (Ghent) 1848–60.

Published and edited by Alexandre Verschaffelt from 1848–March 1850, and thereafter by Ambroise Verschaffelt.

624 plates by B. Leon, A. Lagarde, G. Severeyns, L. Stroobant and P. Stroobant, lithographed by G. Severeyns and L. Stroobant.

REFERENCE BIBLIOGRAPHY

Arnold Arboretum of Harvard University. *Catalogue of the Library.* 2 vols. Cambridge, MA: Harvard University, 1914–17.

Blunt, Wilfrid, and W. T. Stearn. *The Art of Botanical Illustration.* London, 1950.

British Museum (Natural History). *Catalogue of the Books, Manuscripts, Maps and Drawings.* 8 vols. London, 1903–40. Reprint. 1964.

Buchanan, Handasyde. *Nature into Art: A Treasury of Great Natural History Books.* London: Weidenfeld & Nicolson, 1979.

Buchanan, H., and G. Grigson. *Thornton's* Temple of Flora. London, 1951.

Bunyard, E. A. "A Guide to the Literature of Pomology." Royal Horticultural Society. *Journal.* Vol. 40: pp. 414–49. London, 1915.

Calmann, Gerta. *Ehret: Flower Painter Extraordinary.* Boston: New York Graphic Society; London: Phaidon Press Ltd., 1977.

Coats, Alice M. *The Book of Flowers: Four Centuries of Flower Illustration.* London, 1973.

Coats, Alice M. *The Plant Hunters, Being a History of the Horticultural Pioneers, Their Quests and Their Discoveries from the Renaissance to the Twentieth Century.* New York: McGraw-Hill, 1970; London: Studio Vista Limited, 1969.

Daniels, Gilbert S. *Artists from the Royal Botanic Gardens, Kew.* Pittsburgh: Hunt Institute for Botanical Documentation, 1974.

Dunthorne, Gordon. *Flower and Fruit Prints of the Eighteenth and Early Nineteenth Centuries.* (Orig. pub. Washington, D.C., 1938.) Reprint. London: Holland Press, 1970.

Henrey, Blanche. *British Botanical and Horticultural Literature before 1800.* 3 vols. London: Oxford University Press, 1975.

Hutton, Paul and Lawrence Smith. *Flowers in Art from East and West.* London, 1979.

Jackson, Benjamin D. *Guide to the Literature of Botany.* (Orig. pub. London, 1881). Reprint. Koenigstein, 1974.

King, Ronald, ed. The Temple of Flora *by Robert Thornton.* Boston: New York Graphic Society, 1981.

Landwehr, J. H. *Studies in Dutch Books with Coloured Plates Published 1662–1875.* The Hague, 1976.

Lawrence, George, ed. *A Catalogue of Redoutéana.* Pittsburgh: Hunt Institute for Botanical Documentation, 1963.

Lindley Library. *Catalogue of Books, Pamphlets, Manuscripts and Drawings.* London: Royal Horticultural Society, 1927.

Massachusetts Horticultural Society. *A Catalogue of the Library.* Boston, 1918–20.

Massachusetts Horticultural Society. *Dictionary Catalog of the Library.* Boston, 1963.

Nissen, Claus. *Die botanische Buchillustration, ihre Geschichte und Bibliographie.* 2 vols. Second ed. Stuttgart, 1951–52. Another ed. 1966.

Nissen, Claus. *Botanische Prachtwerke, die Blütezeit der Pflanzenillustration von 1740 bis 1840.* Vienna, 1933.

Pierpont Morgan Library. *Flowers in Books and Drawings c 940–1840.* New York, 1980.

Pritzel, G. A. *Thesaurus literaturae botanicae.* Leipzig, 1851. Second ed. Leipzig, 1872. Another ed. Milan, 1950.

Quimby, Jane, and Allan Stephenson. *The Hunt Botanical Catalogue.* 2 vols. Pittsburgh: Hunt Institute for Botanical Documentation, 1961.

Rix, Martyn. *The Art of the Plant World: The Great Botanical Illustrators and Their Work.* Woodstock, NY: Overlook Press, 1980. Published as *The Art of the Botanist,* London: Cameron & Tayleur, 1980.

Royal Botanic Gardens, Kew. *Catalogue of the Library.* London, 1899.

Sotheby's. *The Library of the Stiftung für Botanik, Liechtenstein, Collected by the Late Arpad Plesch.* London: Sotheby's, 1975–76.

Sotheby's. *A Magnificent Collection of Botanical Books . . . [from the Library of] . . . Robert de Belder.* London: Sotheby's, 1987.

Sotheby's. *Pierre-Joseph Redouté's* Les Liliacées: *The Empress Josephine's Copy with the Original Drawings and the Text on Vellum.* London: Sotheby's, 1985.

Stafleu, F. A. and R. S. Cowan. *Taxonomic Literature.* Vols. I–VI. Second ed. Utrecht and Zug, 1976–86.

Stearn, W. T. "Some Books on Camellias." Royal Horticultural Society. *Camellias and Magnolias.* pp. 124–28. London, 1950.

Stearn, W. T. *An Exhibition of Flower Books from the Library of the Society of Herbalists.* London, 1953.

Woodward, B. B. "Redouté's Works." *Journal of Botany.* Vol. 43: pp. 26–30. London, 1905.

PLATE 50

This Amazonian waterlily illustrates the monograph Victoria regia *(Boston 1854) by John F. Allen. Discovered in the 1830s, it took years for botanists to grow this giant in a manmade pond where it could be studied closely. William Sharp made six large chromolithographs after his own paintings of a lily growing in Massachusetts for this landmark American colorplate book.*

Lilium speciosum.

LIST OF THE PLATES

Elizabeth Braun

The List of Plates records each illustration's collaborating artists, printmakers, authors, and publishers. The first line of each description is the title of the print; italic letters identify the source of the title as either the print itself or its text; titles in roman letters signify modern common names. The title and edition given is that of the actual volume from which we photographed the print; following is a description of the print medium, and the names of the artists who collaborated on it when known. A few sentences give additional information on the creation of the print, or about the particular volume from which we photographed it. Locations of original watercolor models for the prints are given when known; current locations are as of July 1989.

All quotes whose sources are not cited are taken from the texts of the works illustrated; in many instances, page numbers are given. Citations noted in parentheses are to be found in the Reference Bibliography, which lists sources for further reading. Definitions of printmaking terms may be found in Buchanan's Appendix. Gavin Bridson's article "The Treatment of Plates in Bibliographical Description" (Society for the Bibliography of Natural History, *Journal*, 7: 4, 1976) was an inspiration for this List, as were many model publications by the Hunt Institute.

1
Double White Camellia; Double Striped Camellia
Etching in aquatint and line by Weddell, after the watercolor by Clara Maria Pope, hand colored with gouache and heightened with gum arabic by the artist.
Curtis, Samuel. *A Monograph of the Genus Camellia*. London 1819. Pl. 3.
Weddell copied Pope's watercolors in aquatint to describe broad, dark leaves and colored petals, with line etching to accentuate their contours. The artist, by serving as colorist, had direct control over the appearance of the finished prints. Pope has used opaque gouache of deep hues to imitate the saturated colors of the foliage. She has covered most of the pigment area with either gum arabic or albumen varnish to imitate the glossiness distinctive of camellia leaves. Several other prints after Pope's flower paintings were published by Samuel Curtis (plates 31 and 32).
Courtesy of New York Public Library, Print Department.

2
Banana flower
Engraving by Jan P. Sluyter, after the watercolor by Maria Sibylla Merian, and hand colored.
Merian, Maria Sibylla. *Dissertatio de . . . insectorum surinamensium*. Amsterdam 1719. Second edition. Pl. 12.
Merian's prints show the stages of development of insects from caterpillar, to pupa, to butterfly or moth. She lived in Surinam, a Dutch colony in South America, to observe these insects and their host plants in the field. Upon her return to Amsterdam, Merian married into the publishing family of that name, had the engravings prepared, and produced her great book. The first edition of her work conveyed sixty prints and appeared in 1705. This edition was edited by her daughter, Johanna Graff, who provided a supplemental twelve plates. This print serves as plate 12 in both editions, and was printed for both from the same plate.
Courtesy of W. Graham Arader III Gallery, New York.

PLATE 51
"Lilium Speciosum" illustrates A Monograph of the Genus Lilium *(London 1877–80). Author Henry Elwes traveled the world over to discover lilies, and cultivated them to perfection in a study garden.*
Walter Fitch, a prolific botanical lithographer, produced some of his best work for this book. Elwes includes a "type specimen" of this species, as well as two varieties.

3
Passionflower
Engraving by Joseph Mulder, after the watercolor by Maria Sibylla Merian, and hand colored.
Merian. *Dissertatio.* Amsterdam 1719. Pl. 21.
Merian's stylization is especially evident in this delicate passionflower, supporting Sitwell's assertion that these prints could serve as embroidery designs. Coloring varies greatly between different examples of the book, many of them perhaps colored by the original owners and not under the supervision of the publisher. An example of this print was seen with the image precisely cut out and laid down on paper of the era; perhaps the work of a woman of leisure.
Courtesy of W. Graham Arader III Gallery, New York.

4
Hellebore, or Christmas rose
Engraving by Nicolas Robert after his own watercolor, and printed at the Imprimerie Royale.
Robert, Nicolas. *Estampes pour servir à l'histoire des plantes.* Paris 1701. Pl. 205.
This work was published several times under various names, and is sometimes listed under the name of the botanist in charge, Denis Dodart. Many of the watercolors were transcribed into line for engraver's models, a procedure which no doubt enhanced the finished print. Robert himself engraved the plates with the help of Abraham Bosse through 1676, and with Louis de Chatillon until 1692. Around 1719 details of flowers and seeds were engraved on the plates; later issues of the prints and a modern issue printed by the Musée Louvre chalcographie carry these details. A special issue of the plates, printed on vellum and hand colored by Henri and Pierre Joseph Redouté and G. Prévost, appeared in the early 1800s, of which the Pierpont Morgan Library owns 171 prints. (Pierpont Morgan Library. *Flowers in Books and Drawings ca. 940–1840.* New York, 1980. Entry 82; pl. 11.)
Courtesy of Dumbarton Oaks Garden Library.

5
April
Engraving by Henry Fletcher, after the design by Pieter Casteels, and hand colored.
Furber, Robert. *The Twelve Months of Flowers.* London 1730. "April" [Pl. 4].
Furber published the names of 450 subscribers on the engraved title page to this calendar. It serves as an advertisement of his inventory of plants and seeds, twenty-seven of which appear in this print alone. Casteels painted the series of twelve watercolors, as well as a complementary series of fruit that Furber published in 1732. *Flowers* appeared in several subsequent editions, with plates engraved on a smaller scale.
Courtesy of George P. Brett, Jr., Collection, Pequot Library, Southport, Connecticut.

6
Magnolia grandiflora
Etching with engraving by Georg Ehret after his own watercolor, and hand-colored by George Edwards.
Catesby, Mark. *The Natural History of Carolina.* London 1754. Second edition. Vol. 2, pl. 61.
Ehret made two of the 220 prints in Catesby's book, both of magnolias. They are distinguished by elegantly inscribed lines that differ greatly from the work of the self-taught author. Magnolias were much sought by English horticulturists who acclimatized them to their estates. Ehret painted this specimen as it bloomed in 1737 at London, traveling across town daily to capture the opening flower. Another magnolia in this work was engraved by Catesby after a design by Ehret, and plainly shows Catesby's labored handling of lines.
Courtesy of a private collection.

7
Aloe tuberosa levis
Engraving in mezzotint with some line, inked in green, pink and gold, with some hand coloring.
Weinmann, Johann. *Phytanthoza iconographia.* Amsterdam 1736. Vol. 1, pl. 66.
The author used drawings by Georg Ehret as models for many of the over 1,000 plates in this work. Few are credited to Ehret, and similarly little credit is given to the engravers of this pioneering work of botanical prints engraved to be inked in color. Mezzotint imitates the smooth surfaces of this succulent plant, an effect for which line etching or engraving is unsuited. This example of the complete work is distinguished by delicate and detailed early impressions from the plates. Ehret's biographer, Calmann (1977, p. 112) notes that succulents appear frequently in his oeuvre; twenty-two such drawings were acquired by the Royal Society of London in 1737. Ehret frequently created similar watercolors for different patrons from his stock of working drawings.
Courtesy of W. Graham Arader III Gallery, New York.

8
Cereus cactus
Etching by Johann Jakob Haid, after the watercolor by Georg Ehret, and hand colored.
Trew, Christoph. *Uitgezochte planten.* Amsterdam 1771. (Dutch edition of *Plantae selectae*, Augsburg 1750–73.) Pl. 31.
Trew acted as Ehret's most important patron, commissioning hundreds of detailed botanical watercolors over several decades. Many of these were published by Trew in this book and in *Hortus nitidissimis* (plate 9) which also contained the work of other artists. *Plantae selectae* features exotic plants such as magnolias, bananas, and pineapples in several plates each. Ehret painted these and the subtropical cereus at London, where gardeners recreated the plants' natural habitats by heating glass houses with stoves. Ehret's accomplished preliminary drawings for this work are at the British Museum (Natural History), London. Trew's collection of watercolors now belongs to Erlangen University, Germany.
Courtesy of New York Public Library.

9

Corona Imperialis William Rex Crown imperial
Etching by Johann Seligmann, after the watercolor painted by J. C. Keller in 1757, and hand colored.
Trew, Christoph. *Hortus nitidissimis.* Nuremberg 1750–86. Pl. 40.
Issued concurrently with *Plantae selectae*, this book pictured garden flowers, perhaps to complement *Plantae*'s exotic species. Trew tapped his collection of hundreds of botanical watercolors by Ehret and local artists for the prints. Ehret contributed forty-four drawings, indicated by "Eh" or his full signature in a bottom corner of the print. The drawings are boldly and accurately colored, and exhibit great detail.
Courtesy of Donald Heald Fine Art, New York.

10

Iris Vulgaris Germanica
Engraving with etching by J. Fougeron, after the watercolor by John Edwards, and hand colored.
Edwards, John. *The British Herbal.* London 1770. Pl. 1.
Fougeron's great skill is evidenced by the exquisite laying of lines on the buds and on the bearded "falls," or lower petals of this iris. The style of engraving is finer than that of Trew's contemporary works (plates 8 and 9), and the hand coloring is transparent, allowing the lines to shine through to delineate anatomical details. The whole was reissued in 1775 as *A Select Collection of 100 Plates. . . .* Many years later Edwards produced an even finer flower book with prints completely by his own hand (plates 15 and 16).
Courtesy of the Library of the New York Botanical Garden, Bronx, New York.

11

Helianthus
Etching by John Sebastian Miller after his own design, inked in brown, and hand colored.
Miller (Müller, or Mueller), John S. *Illustratio systematis sexualis linnaei.* London 1770–7. Class 19, Order 3.
In this issue of the book, each of the 104 plates appears printed in black without coloring, and printed in the present brown or like hue with hand coloring. Miller has chosen his plants to typify different orders of the Linnean system of classification. This was determined by the parts of flowers, female and male, which appear in different proportions throughout the plant kingdom. He figures a representative part of this composite flower dissected along the bottom of the plate. The brown ink carried by the plate blends with the golden hand coloring of the petals, in a convincing simulation of the actual flower.
Courtesy of the Library of the New York Botanical Garden, Bronx, New York.

12

Leontodon taraxacum Dandelion
Etching with engraving by William Kilburn after his own design, and hand colored.
Curtis, William. *Flora londinensis.* London 1775–98.
Many artists and printmakers and about thirty colorists worked on the 432 lifesize renderings of field flowers near London. Many sets were sold uncolored at a reduced cost; details in the text descriptions enabled the purchaser to color them with some fidelity. The prints were issued unnumbered in parts of six each with corresponding text, and subscribers thus ordered them according to Linnaeus's or their preferred classification system.
Courtesy of New York Public Library.

13

La Pivoine commune Common peony
Etching, hand colored.
Buc'hoz, Pierre. *Collection précieuse.* Paris [1776]. Pl. 49.
The author issued hundreds of botanical etchings, though only rarely giving credit to artists and printmakers. This work is divided into two "centuries" of 100 plates each. The first century is notable for its reproduction of Chinese watercolors, and the second conveys familiar garden flowers. The second century contains prints gleaned from other works by the author; it is numbered 1–CC but contains only 100 prints.
Courtesy of the Library of the New York Botanical Garden, Bronx, New York.

14

L'Amaryllis d'Eté
Etching, hand colored.
Buc'hoz. *Collection.* Paris [1776]. Pl. 63.
This volume of flower prints had no text, and the images are not carefully drawn. They appear to have been produced in order to amuse—witness the charming two-legged butterfly creature. The gold borders are reminiscent of those placed by painters of *vélins* around their careful specimen drawings at the Museum d'histoire naturelle. Buc'hoz published prints of many of these fine watercolors in his *Histoire universelle du règne végetale* (1773–8).
Courtesy of the Library of the New York Botanical Garden, Bronx, New York.

15

Large Orange Lilly
Etching by John Edwards after his own design, inked in pale ink and printed in 1780, and hand colored.
Edwards, John. *A Collection of Flowers Drawn after Nature.* London 1801.
Edwards's treatment of flowers conforms to his own style and high standard of printmaking, quite out of the norm of English botanical printmaking. Some of the prints employ softground etching. This produces soft planes of tones, and fuzzy, soft-edged lines.
Courtesy of the Minnich Collection, Minneapolis Institute of Arts.

16

Various Tulips
Etching by John Edwards after his own design, inked in pale ink, and hand colored.

Edwards. *A Collection of Flowers*. London 1801.
The artist/author concerns himself with garden flowers in this portfolio of seventy-seven etchings. The delicate etching, inked in pale hues, does not interfere with the careful hand coloring. This work is extremely rare in North America, and we have located only one complete example at the Metropolitan Museum of Art, and a large fragment at the Minneapolis Institute of Arts, which happily was able to furnish this and the previous plate.
Courtesy of the Minnich Collection, Minneapolis Institute of Arts.

17
Tulipe des jardins
Stipple etching by Pierre François Le Grand, after the watercolor by Gerrit van Spaendonck, and hand colored.
Spaendonck, Gerrit van. *Fleurs dessinés d'après nature*. Paris 1801.
Stipple prints illustrate the finest French botanical books in the first decades of the nineteenth century. Van Spaendonck, the painter of flower *vélins* at the Jardin des plantes, was the first to apply it to a complete work of folio prints. The printmakers created thousands of minute concavities in the plate to hold ink, through a blend of engraving and etching. A special, downcurved burin was used to peck a plate covered with ground, in order to expose the metal. The plate was then immersed in acid; light tones could be covered with ground and the plate immersed yet again to create the darker tones. At times, the pecking action created deep concavities that made etching unnecessary, and the tiny dots of this type of stipple print are clear-edged and triangular. This work was issued inked in black; some, including this one, were also inked in color.
Courtesy of the Minnich Collection, Minneapolis Institute of Arts.

18
Erica Sebana Heath
Engraving in line with some stipple by Daniel Mackenzie, after the watercolor by Franz Andreas Bauer, and hand colored.
Bauer, Franz. *Delineations of Exotick Plants*. London 1796.
Through the patronage of Sir Joseph Hooker, Bauer was installed as artist-in-residence at Kew Gardens, a position that still exits today. The artist illustrated many fine English botanical books, and produced two fine folios of illustrations of his own. This work conveys thirty beautiful prints, of which twenty show heaths collected in South Africa. Although they represent species new to science, no attendant text was published. The Preface asserts that "each figure is intended to answer itself every question a Botanist can wish to ask" concerning plant anatomy, and the prints included magnified dissections of the flower to permit Linnean classification. The Hunt Institute has some of Bauer's watercolor models for this work.
Courtesy of the Library of the Hunt Institute for Botanical Documentation, Pittsburgh.

19
A Group of Tulips
Engraving in mezzotint and line made by R. Earlom in 1798, after the painting by Philip Reinagle, inked in several colors, and hand colored.
Thornton, Robert. *The Temple of Flora*. London 1799–1807.
Miller explicated the Linnean system of classification by illustrating species typical of its different orders, and executing large line etchings with floral dissections (plate 11). Thornton's introduction to *The Temple of Flora* expressed the same goal, but the plates celebrate the flowers, rather than tutoring the reader in Linnaeus's system. Our print, a "proof before letters," shows the engraver's name in the image vignette at lower right and not, as in the final state, engraved in the blank below the image. The volume from which this and the following two plates were taken was owned by the great bibliophile Estelle Doheny. (This image also appears on our dustjacket.)
Courtesy of W. Graham Arader III Gallery, New York.

20
A Group of Auriculas
Etching in aquatint, stipple and line made by F. C. Lewis and J. Hopwood in 1803 after the painting by Peter Henderson, inked in three colors, hand colored and with gum arabic highlights.
Thornton. *The Temple of Flora*. London 1799–1807.
Dunthorne (1938) provides the standard accounting of the different states of issue of the thirty-two prints in Thornton's *Temple*, erected to the legacy of Linnaeus. These auriculas seem to represent the second state of the print, lacking three eagles which soar above the mountains in the first state.
Courtesy of W. Graham Arader III Gallery, New York.

21
The Sacred Egyptian Bean
Aquatint etching with stipple made by Burke and F. C. Lewis in 1804, after the painting by Peter Henderson, inked in five colors and hand colored.
Thornton. *The Temple of Flora*. London 1799–1807.
The memorable print shows color printing to its best advantage. It supports Ronald King's assertion that Thornton's prints exploit "all that is most sensational in nature." The articulation of veins in the leaves and petals was achieved by covering the light areas with ground and immersing the plate in acid, repeating this procedure to arrive at the darkest tones.
Courtesy of W. Graham Arader III Gallery, New York.

22
Iris germanica
Etching in stipple by De Gouy, after the watercolor by Pierre Joseph Redouté, inked *à la poupée* in colors and printed under the direction of Tassaert at the press of Langlois, and hand colored.
Redouté, Pierre Joseph. *Les liliacées*. Paris 1802–16. Vol. 6, pl. 309.

This and plates 23 and 24 have been photographed from an example of the work conserved by Judy Reed, Conservator to the Library of the New York Botanical Garden. Acidity in the paper, one of the great destructive agents of old books, was neutralized. Mold which had developed at some damp period in the book's history was removed. Each leaf was delicately hinged into a new binding of acid-free materials. The boxed volumes are now kept in the Library's humidity and temperature controlled stacks.
Courtesy of the Library of the New York Botanical Garden, Bronx, New York.

23
Lilium superbum
Etching in stipple by De Gouy, after the watercolor by Redouté, inked *à la poupée* in five colors and printed under the direction of Tassaert at the press of Langlois, and hand colored.
Redouté. *Les liliacées.* Paris 1802–16. Vol. 3, pl. 103.
The artist maintained an atelier of eighteen stipple etchers who produced the plates for his many colorplate books. By coincidence, De Gouy made the three remarkable prints we feature from *Les liliacées.* His name is engraved differently on each print, for a specialist in engraving letters added inscriptions after the etchers finished the image. Jean Tassaert, who supervised the creation and printing of the plates, was a professor at the Ecole Polytechnique at Paris.
Courtesy of the Library of the New York Botanical Garden, Bronx, New York.

24
Agapanthus umbellatus
Etching in stipple by De Gouy, after the watercolor by Redouté, inked *à la poupée* in four colors and printed under the direction of Tassaert at the press of Langlois, and hand colored.
Redouté. *Les liliacées.* Paris 1802–16. Vol. 1, pl. 4 [actually pl. 6].
Redouté painted the watercolor models for these prints on vellum, a finely textured surface suited for delicate brushwork. Traces of his underlying pencil sketches are visible, and he has signed many of them. Sotheby's auctioned 468 of the paintings in 1985 and the catalogue illustrates each with good color reproductions. Arader bought the collection and most of it has been dispersed. A number of the paintings have been donated to the Library of the New York Botanical Garden. This erroneously numbered plate has been bound in its correct location with the corresponding text identifying it as plate 6 (see plate 20).
Courtesy of the Library of the New York Botanical Garden, Bronx, New York.

25
Silver-Rock Mellon
Etching in aquatint with some line made in 1812, after the painting by George Brookshaw, inked in four colors, and hand colored.
Brookshaw, George. *Pomona britannica.* London 1804–12. Pl. 67.
The ninety plates of this work are large and powerful images due to their dark squared vignettes. This book may embody the English horticulturist's answer to the florist's wonders printed by Thornton (plates 19–21) and Curtis (plates 31 and 32). This melon is one of several and is well complemented by a similar series of very impressive pineapples. The texture of its skin is completely described in carefully worked aquatint, as are its pulp and seeds; color printing heightens its realism.
Courtesy of W. Graham Arader III Gallery, New York.

26
Peaches
Etching in aquatint with some line made in 1806, after the painting by George Brookshaw, inked in three colors, and hand colored.
Brookshaw. *Pomona britannica.* London 1804–12. Pl. 26.
Although the melon print of plate 25 has a carefully engraved tablet and publisher's line at its base, these peaches have none. Both forms are found throughout the work, and neither indicates a different state of impression or edition of the prints. The appeal of these prints owes much to the velvety texture of the aquatint, which is at its best in its first impressions; later pulls have a distracting, uneven texture.
Courtesy of W. Graham Arader III Gallery, New York.

27
Alcea rosea
Etching in stipple made by Alexandre Clement in 1808, after the watercolor by Pancrace Bessa, inked *à la poupée* in four colors.
Bessa, Pancrace. *Fleurs et fruits.* Paris 1808.
Like Redouté a pupil of van Spaendonck (plate 17), Bessa painted in much the same style and also used stipple etching to translate his delicate watercolors into prints. There is no hand coloring on this print yet it stands as a successful, complete image. The illusion of three dimensions in the flowers and leaves is achieved by inking in both light and dark hues of red and green. The stippling dots are placed very close together, and the lack of hand coloring allows the clean white paper to show through; the result is a fresh, brilliant image.
Courtesy of the Library of the Hunt Institute for Botanical Documentation, Pittsburgh.

28
Poppies
Etching in stipple by Louis Charles Ruotte, after the watercolor by Jean Louis Prévost, inked *à la poupée* in seven colors and printed at the press of Vilquin, and hand colored.
Prévost, Jean Louis. *Collection des fleurs et fruits.* Paris 1805. Pt. 11, pl. 42.
Prévost, another contemporary of Redouté, painted in much the same style. Like van Spaendonck (plate 17) and Bessa (plate 27), he prepared a large folio of color prints after his

paintings. The manner of etching is more linear, indicating the use of roulette, and it is also coarser. The larger dots of intensely colored ink result in a particularly strong image—aided of course by the artist's graceful draftsmanship. The motif of moutan peonies seen nodding from behind is repeated throughout French art of this era.
Courtesy of Donald Heald Fine Art, New York.

29
Hollyhock
Etching in stipple by Ruotte, after the watercolor by Prévost, inked *à la poupée* in two colors and printed at the press of Vilquin, and hand colored.
Prévost. *Collection.* Paris 1805. Pt. 4, pl. 15.
Prévost designed this portfolio to supply decorative motifs to artisans in other crafts, especially those designing textiles. Other plates feature fruit still lifes and bouquets carefully inked in several colors. The plates were issued in "Cahiers," or parts, which, conveniently for the bibliographer, are indicated at top.
Courtesy of Donald Heald Fine Art, New York.

30
Amaryllis
Etching in line by Weber, after the watercolor by Strenzel, and hand colored.
Trattinick, Leopold. *Thesaurus botanicus.* Vienna 1805–19. Pl. 41.
While stipple etching reigned in France, German flower books were mainly illustrated with etching and engraving in line. This amaryllis exhibits the most careful hand coloring; the colorist has created a delicate network of veins in the petals without the aid of etched guidelines. However skillful, this style of illustration looks much like prints of the eighteenth century.
Courtesy of W. Graham Arader III Gallery, New York.

31
Anemones
Etching in aquatint by Weddell, after the painting by Clara Maria Pope, printed in pale ink, and hand colored.
Curtis, Samuel. *The Beauties of Flora.* Gamston 1806–20.
Curtis was well known for his garden journal when he undertook this fine folio of ten flower aquatints. The fame of the artist, Clara Pope, a member of the Royal Academy, no doubt encouraged him in this venture. The form of the prints imitates the squared, scenic vignettes of Thornton's slightly earlier work (plates 19–21).
Courtesy of New York Public Library, Print Department.

32
Dianthus
Etching in aquatint and line by Weddell, after the painting by Clara Maria Pope, and hand colored.
Curtis. *The Beauties of Flora.* Gamston 1806–20.
Whereas the similar work by Thornton (plates 19–21) focused on exotic flowers, Curtis, a florist, featured familiar varieties. The work was apparently issued as a collection of decorative prints without text, and is now very rare. Curtis, Pope, and Weddell also collaborated on *A Monograph of the Genus Camellia* (see our plate 1).
Courtesy of New York Public Library, Print Department.

33
Abricot-pêche Apricot
Etching in stipple with some line by Louis Bouquet, after the watercolor by Pierre Turpin, inked in five colors and printed at the press of Langlois, and hand colored.
Duhamel du Monceau, Henri. *Traité des arbres fruitiers.* Paris 1807–35. Pl. 104.
The first edition consisted of 180 engraved plates not intended for coloring and was published in 1768. Turpin and Pierre Poiteau, botanical artists, teamed together to illustrate a new edition of Duhamel du Monceau's *Traité* long after the author's passing. Each print shows the branch with fruit and leaves, a flower, a split fruit showing seeds, and occasionally other details. Poiteau alone issued impressions from the plates in 1846, as *Pomologie française*, easily distinguished by the omission of Turpin's name at lower left. The 1800–25 edition of Duhamel du Monceau's *Traité des arbres et arbustes* carries equally exquisite plates of shrubs and trees.
Courtesy of Donald Heald Fine Art, New York.

34
Fico dell' Osso Bear fig
Etching in aquatint, stipple and line by Giuseppe Pera, after the painting made by Domenico del Pino in 1826, inked in five colors, and hand colored.
Gallesio, Giorgio. *Pomona italiana.* Pisa 1817–39. Vol. 2.
A multitude of artists and printmakers, all Italian, collaborated on this Italian rival to Poiteau's edition of Duhamel du Monceau (plate 33). Distinctive of this work is its discreet and skillful use of aquatint, here especially evident on the bird. The prints are larger than those of the French fruit book, and their inscriptions are quite different. These accomplished Italian printmakers have surely made their mark elsewhere in decorative prints and illustration, if not in the field of botany.
Courtesy of W. Graham Arader III Gallery, New York.

35
Pinus Pinaster Pine
Engraving in line with stipple and roulette shading by Daniel Mackenzie, after the watercolor by Georg Ehret, inked in colors, and hand colored.
Lambert, Aylmer. *A Description of the Genus Pinus.* London 1837. Third edition. Vol. 1, pl. 10.
The majority of the watercolor models for this work's forty-two prints were painted by Ferdinand Bauer, with only this one contributed by the great Ehret. The artist painted it about 1744, and Lambert borrowed it from Sir Joseph Banks around 1800 to serve as an engraver's model. It is now preserved in the British Museum (Natural History). Mackenzie created the plate which prints the convincing illusion

of three-dimensional pine needles. Mackenzie had previously engraved plates after the paintings of Ferdinand's brother Franz, for the latter's *Delineations* (plate 18). The first edition of this work was issued in only twenty-five copies, this image serving as plate 5.
Courtesy of New York Public Library.

36
Passiflore quadrangularis Passionflower
Etching in roulette and line by Robert, after the watercolor by Pierre Poiteau, etched, inked in blue, and printed under the direction of P. Duménil, and hand colored.
Tussac, F. Richard de. *Flore des Antilles*. Paris 1808–27. Vol. 4, pl. 10.
Poiteau collected tropical plants in the West Indies for the Jardin des plantes at Paris. The author himself was an artist, but his drawings were destroyed before publication. Poiteau, drawing upon de Tussac's herbarium and aided by his field experience, rendered most of the watercolor models for the plates. The blue colored ink is a harmonious undertone to the hand coloring, which varies from lavender to green.
Courtesy of W. Graham Arader III Gallery, New York.

37
Clusia rosea Clusia
Etching in stipple and roulette with some line by Wouillaume, after the watercolor by Poiteau, etched, inked in four colors, and printed under the direction of P. Duménil, and hand colored.
Tussac, *Flore des Antilles*. Paris 1808–27. Vol. 4, pl. 15.
Several printmakers worked to illustrate about 140 tropical, mostly West Indian plants, for this fine folio. Most of the prints are stipple, with roulette used for shading and details; the manner of stippling is quite different from that used by Bouquet for the apricot (plate 33) in a contemporary work. Stippling could be fine or coarse, depending on the tools used to mark the tiny dots, and the length of time each area of dots was exposed to acid. The flat depiction of the leaves is reminiscent of herbarium specimens, in which plants are arranged between sheets of heavy paper and pressed. The result is an oddly one-dimensional image. Merian's Passionflower (plate 3) is another example of this style.
Courtesy of W. Graham Arader III Gallery, New York.

38
Paeonia Moutan. Var. b. Moutan peony
Etching in stipple by Louis Bouquet, after the watercolor by Pierre Joseph Redouté, inked *à la poupée* in four colors and printed at the press of Langlois, and finished by hand with watercolor.
Bonpland, Aimé. *Description des plantes rares cultivées à Malmaison et à Navarre*. Paris 1813. Vol. 1, pl. 23.
Bonpland and Alexander Humboldt traveled throughout South America as naturalists, recording natural phenomena. Bonpland collected and drew plants, and his achievement earned him the directorship of Empress Josephine's great gardens at the estate Malmaison. *Description* illustrates a selection of plants painted by Redouté and Bessa. They are more beautiful than scientific, and the quality of stipple etching by Bouquet is magnificent. Here the printer has carefully mixed blue among the red petals, enhancing the depth of the flower. A comparable volume by Étienne Ventenat, *Jardin de la Malmaison* (1803–5), is more botanically inclined. Redouté painted almost 200 detailed and graceful watercolors for these works, even as he produced the 486 paintings for *Les liliacées* (plates 22–24) and engaged in other work.
Courtesy of the Library of the New York Botanical Garden, Bronx, New York.

39
Antirrhinum latifolium
Etching in stipple by F. W. Meyer, after the watercolor by G. W. Voelcker, inked *à la poupée* in five colors, and finished by hand with watercolor.
Hoffmannsegg, Johann. *Flore portugaise*. Berlin 1809–40. Pl. 50.
Stipple etchings dominated French botanical illustration, and were preferred for the illustration of government expedition reports. Studios at Paris maintained groups of printmakers specializing in the medium and its attendant color inking. Gallesio's *Pomona italiana* (plate 34), produced in Pisa, and the comprehensive flora by Hoffmannsegg produced in Berlin, are isolated yet masterful embodiments of the stippler's art. Almost all of the latter's prints are based on watercolors by G. W. Voelcker, and many printmakers, all German, were involved. The structure of the book is a series of plates showing allied species; thus, all the antirrhinums are placed in a group, with our illustration being the most lovely.
Courtesy of the Library of the New York Botanical Garden, Bronx, New York.

40
Campanula peregrina
Etching in stipple with engraving in roulette and line by F. W. Meyer, after the watercolor by G. W. Voelcker, inked *à la poupée* in six colors, and finished by hand with watercolor.
Hoffmannsegg. *Flore portugaise*. Berlin 1809–40. Pl. 83.
The 114 plates picture the wild progenitors of the grander, cultivated varieties of flowers native to Portugal. Many of the images, such as this campanula, show the entire plant in a seemingly three-dimensional portrait. With great naturalism, the physiological details typical of dying leaves, stunted shoots, and withering flowers are included. Therefore, a less than pristine example found in the field (which is likely) could be easily identified. The printmaker effected wonders with his tools in order to create the illusion of the shoots in back of the main stalk angling away, and the inker has used a different color of green to enhance this realism.
Courtesy of the Library of the New York Botanical Garden, Bronx, New York.

41
Rosa Indica Cruenta
Etching in stipple by Langlois, after the watercolor by Pierre Joseph Redouté, inked *à la poupée* in four colors and printed at the press of Rémond, and finished by hand with watercolor by the artist.
Redouté, Pierre Joseph. *Les roses*. Paris 1817. Vol. 1, opposite page 123.
This is the most famous of Redouté's many flower books. A plethora of modern reproductions attest to its enduring beauty and popularity. It is hard to imagine a flower picture approaching the grace of this particular image. Empress Josephine favored Redouté with patronage that guaranteed the success of *Les liliacées* (plates 22–24), and he dedicated this monograph on her favorite flowers to her memory. This and plate 42 were photographed from a large-paper copy once the property of the great bibliophile Estelle Doheny. The watercolor, painted on vellum, currently belongs to Arader.
Courtesy of Donald Heald Fine Art, New York.

42
Rosa Gallica Versicolor
Etching in stipple by Langlois, after the watercolor by Redouté, inked *à la poupée* in four colors and printed at the press of Rémond, and finished by hand with watercolor by the artist.
Redouté. *Les roses*. Paris 1817. Vol. 1, opposite page 135.
The "versicolor" in this rose's name describes the lovely, random streaks of pink and darker pink upon its petals. Whereas hand coloring would introduce either the intrusive outlines for these streaks, or else the inaccuracy of freehand work, color inking ensures their naturalism, and promotes consistency among the impressions taken from the plate. The minimal hand coloring on this plate was applied by the artist himself, as this is a special, large-paper copy of the work. The flowers pictured in the 169 prints vary from stupendous cabbage roses to briars with tiny, simple flowers along their stems. A few were adapted by the artist for inclusion in his later *Choix* (plates 44 and 45).
Courtesy of Donald Heald Fine Art, New York.

43
Amaryllis crocata
Etching in aquatint with some line by Robert Havell, Jr., after the watercolor by Priscilla Susan (Mrs. Edward) Bury, inked and printed in three colors, hand colored and highlighted with gum arabic at the press of Havell.
Bury, Priscilla. *A Selection of Hexandrian Plants*. London 1831–34. Pl. 16.
Printmaker Robert Havell is best known to naturalists through the 435 huge illustrations to Audubon's *Birds of America* (1827–38). His deft handling of fine-grained aquatint secured his employment for costly, luxurious works such as Mrs. Bury's. Havell created various small folios of prints such as this even as he completed Audubon's monumental work; Audubon evidently approved of this diversion of Havell's labors, for he subscribed to *Selection*. Mrs. Bury apparently secured her patrons before engaging Havell, and her images are lovely and personalized portraits of flowers and butterflies. Her motivation to paint was the desire to preserve their "fugitive beauties," and Havell has enhanced their beauty by careful inking in pigments sympathetic to the hand coloring.
Courtesy of W. Graham Arader III Gallery, New York.

44
Gentiane sans tige [sic] *Pivoine officinale à fleurs simples* Peony
Etching in stipple, after the watercolor by Pierre Joseph Redouté, etched, inked *à la poupée* in two colors and printed at the press of Langlois, and hand colored by the artist.
Redouté, Pierre Joseph. *Choix des plus belles fleurs*. Paris 1827–33.
This print was photographed from a large-paper copy of the *Choix* comprised of early pulls from the plates printed on especially high-quality paper. Redouté himself is said to have finished the prints of this state of issue with hand coloring. The engraved titles on many of these first pulls were incorrect, and this one has been corrected by a previous owner in pencil. An index issued after the plates rectified the mistakes, and the engraved inscriptions have been similarly corrected on later impressions. This example of the book carries the inkstamp of Tsarskoe Selo, the Romanoff's summer palace outside of Petrograd (Leningrad), and must have been treasured by members of the royal family.
Courtesy of Donald Heald Fine Art, New York.

45
Hortensia Hydrangea
Etching in stipple, after the watercolor by Redouté, etched, inked *à la poupée* in two colors and printed at the press of Langlois, and finished by hand with watercolor by the artist.
Redouté. *Choix*. Paris 1827–33.
Antique prints often crumble with time, because no papermaker could predict the aging process of his product. High humidity and temperature enhance the destructiveness of acidic fibers often found in the paper itself. Redouté's *Choix* used paper which often shows the typical signs of aging: chipping edges, brittleness, and darkening of the creamy color of the paper. The volume we photographed for this and the previous plate is a special, large-format issue of the work with paper not prone to age in this manner. The delicate pink of this hydrangea glows on the white paper, whereas even slightly darkened paper would have destroyed the effect.
Courtesy of Donald Heald Fine Art, New York.

46
Pictorial title page
Calligraphy designed and engraved by Tomkins, the remainder etched and engraved by R. Williamson, perhaps after his own design, and hand colored.
Sibthorp, John. *Flora graeca*. London 1806–40. Vol. 8, title page.

Almost 1,000 engraved and etched plates illustrate this comprehensive work on the wildflowers of Greece. The plates are drawn in the manner of herbarium leaves, even though artist Ferdinand Bauer accompanied Sibthorp in his observation of plants in the field. The leaves and flowers are drawn flat on the page with no illusion of three dimensions. This leads to some stylization and often the graceful draping of stems across the print, especially in the case of vines. Each of the volumes has its own pictorial title page, with vignettes of Grecian scenery. It is notable that the calligraphic designer and engraver Tomkins is given credit with an inscribed line.
Courtesy of the Library of the New York Botanical Garden, Bronx, New York.

47
Stanhopea tigrina Orchid
Lithograph drawn by M. Gauci, after the watercolor by Mrs. Withers, colored by hand and highlighted with gum arabic.
Bateman, James. *The Orchidaceae of Mexico and Guatemala.* London 1837–43. Pl. 37.
Sibthorp's flora (plate 46) was a champion of intaglio printmaking in botanical works, even as publishers switched to lithography as a cheaper method of illustration. In the 1840s, lithography took the field, and its relatively inexpensive and less complex production requirements facilitated the copious illustration of botanical books, and the proliferation of more popular, less expensive publications. Bateman used lithography for *Orchidaceae* to produce enormous, accurate, hand-colored prints of orchids. The book's appearance signals a shift in attention from the cultivated beauty of garden flowers to these naturally beautiful tropical species. Orchidomania seized the English of the Victorian era much as tulipomania had captured the Dutch 200 years earlier. The lifesize watercolor model for this print is owned by the Hunt Institute.
Courtesy of the Library of the New York Botanical Garden, Bronx, New York.

48
Epidendrum vitellinum Orchid
Lithograph drawn by M. Gauci, after the watercolor by Miss Drake, printed by Paul Gauci, colored by hand and highlighted with gum arabic.
Lindley, John. *Sertum orchidaceum.* London 1837–41. Pl. 45.
William Curtis began publishing the *Botanical Magazine* just before 1800 to relate descriptions, scientific names, and illustrations of new species to amateurs and botanists. The work disseminated information as well as botanical motifs used by decorative artists. Lindley intended his monograph on orchids to bring fifty new varieties to the public, in a more luxurious format than Curtis's work. This lovely orchid was acquired by Theodore Hartweg, a collector active in South America who brought many new orchids to science. It had already been illustrated in Curtis's *Botanical Magazine*, and Lindley justified featuring it again by denigrating its illustration in the former work. Miss Drake painted this plant from a prize specimen pampered to perfection in the author's own hothouse.
Courtesy of New York Public Library.

49
Rhododendron fulgens
Lithograph drawn by Walter Hood Fitch after his own design based on preliminary sketches by Joseph Dalton Hooker, printed at the press of Vincent Brooks, hand colored and highlighted with gum arabic.
Hooker, Joseph. *The Rhododendrons of Sikkim-Himalaya.* London 1849–51. Pl. 25.
During the botanical exploration of foreign regions, the English showed special interest in woody shrubs that could be acclimatized to their weather conditions. Mark Catesby (plate 6) helped finance his travels in America by sending shoots home to patrons, and the publication of his *Natural History of Carolina* popularized the American shrubs available from English nurserymen. Hooker hazarded a wild journey to the Himalayas, on the fringe of English colonization in Asia, and discovered many species. Hooker's father supervised the publication of this monograph from his son's notes and sketches, sent directly from the field as the expedition continued. The artist Walter Fitch perfected the explorer's drawings on stone.
Courtesy of the Library of the New York Botanical Garden, Bronx, New York.

50
Victoria regia Waterlily
Chromolithograph drawn by William Sharp, after his own painting, printed in five colors by Sharp & Son.
Allen, John F. *Victoria regia.* Boston 1854.
This magnificent waterlily was discovered by plant collector Eduard Poeppig in the 1830s in the Amazon basin. The ribbing structure of its enormous floating leaves is said to have inspired the engineer Joseph Paxton in his design of the Crystal Palace at London in 1851. William Sharp painted the models for this portfolio of six prints from a plant growing near Boston. It is notable among American flower books for the size of its prints, and for its status as a pioneer publication in the new medium of chromolithography. In this method of printing, colors were applied from different stones. Four of these color overlays were printed over a base drawing in black; secondary colors such as green were produced where colors overlapped. Thus a print could be colored with more speed and precision than traditional methods of hand coloring allowed. Chromolithography developed in sophistication, and this early American effort can be compared to the later work of plate 52. A similar folio showing the plant at water level was produced at London by William Hooker in 1847, with four prints by Walter Fitch.
Courtesy of Hirschl & Adler Gallery, New York.

51
Lilium speciosum
Lithograph drawn by Walter Hood Fitch after his own design, and hand colored.

Elwes, Henry. *A Monograph of the Genus Lilium.* London 1877–80.

It is estimated that Fitch drew around 10,000 botanical lithographs throughout his long career. Much of his work comprised the illustration of modest volumes and periodicals, with some notable exceptions being the folio illustrations to Hooker's works on Himalayan flora (plate 49) and Lindley's monograph. His style of drawing is outline that conveys little shading, which is left to the hand colorist. Fitch drew this lily from a type specimen growing in Lindley's garden, and includes two varieties flanking it. Lindley developed a passion for lilies and traveled the world over to collect them; finding no reference work to facilitate his identification of new species, he wrote this volume.

Courtesy of Donald Heald Fine Art, New York.

52

Cattleya percivaliana Orchid

Chromolithograph printed from several stones, after the design by Henry Moon.

Sander, Henry. *Reichenbachia.* St. Albans 1886–94. Vol. 1, pl. 2.

This chromolithograph typifies the beautiful prints produced at century's end for amateur cultivators of exotics. The text accompanying each print relates its appropriate care, and a history of the variety. Sander came from a family of orchid importers and growers, and popularized the passion for a seemingly inexhaustible supply of different varieties. This print was photographed from a special issue of the work rendered almost immovable by its enormous format, richly embossed binding, and heavy paper. Chromolithography, like the stipple prints inked *à la poupée* at the beginning of the century, has no distracting outlines typical of line etchings and engravings. The form of the flower is built of colored pigments, in a convincing facsimile of Moon's original watercolor. Many other periodicals catered to the peculiar interest orchids aroused, such as Warner's *Orchid Album.* The watercolor models are at times on the market, and Arader currently has a few hundred of them. Individual hothouse horticulturists commissioned artists to portray their prize flowers and kept private collections of watercolors; it was an artistic niche often filled by women.

Courtesy of the Library of the New York Botanical Garden, Bronx, New York.

APPENDIX TO THE NEW EDITION

Handasyde Buchanan

PRINTING TECHNIQUES

NATURAL HISTORY PLATES have been made by a variety of techniques through the centuries, and it is not possible to appreciate their finer points without knowing something of the methods involved and the results that could be achieved. The great age of natural history books can be divided into three periods:

Its Rise, 1700–80 The first books with coloured illustrations began to be produced about 1700. Copper engraving, hand-coloured, was the medium of illustration.

The Grand Era, 1780–1830 The copper engravings continued, but the most important works were illustrated with stipple engravings in France, and with aquatint or mezzotint in Britain. The plates of this period were partly printed in colours, partly coloured by hand.

The Gradual Decline, 1830–60 and later This was the age of the lithograph, originally an interesting method with hand-coloured plates, but one that became ever cheaper, and finally, after the arrival of chromolithography, sometimes very nasty indeed.

TECHNIQUES BEFORE 1700

The wood-cut, first developed around 1400, was the earliest method of all. It had ceased to be used on a large scale by 1700, when our period starts, but it is perhaps worth a mention to point out the basic principles of printing.

The original picture was first copied on to a wood block. Then the space between the drawn lines was cut away with a sharp knife, so that these lines, standing proud of the rest of the block, would receive the ink and in due course print black. The image had to be drawn and carved in reverse, of course, as with all the methods of printing discussed here, so that when the inked block was pressed on to a sheet of paper the print made by it would be the right way round.

Wood engraving is a refinement of the wood-cut, in which the engraver uses a burin, the fine steel cutting tool of copper engravers, to obtain a multitude of fine lines that result in subtle gradations of grey tones.

COPPER ENGRAVING

This method was used throughout the eighteenth century and into the early nineteenth, and a large number of very fine plates were produced in this way. The original drawing was transferred on to a copper plate in reverse, usually by tracing or similar means. Then the lines were cut away with a burin—

the deeper the cut, the more ink it takes up and the blacker is the eventual printed line. The plate was then inked and the print was taken from it.

Copper engraving, like the wood-cut and wood engraving, was basically a lineal method and did not allow for very much in the way of light and shade. The following methods enabled more sophisticated effects to be achieved.

MEZZOTINT

The mezzotint was invented in Germany in the seventeenth century. To prepare a copper plate for mezzotint printing it was first roughened all over its surface so that it would, if left like this, print completely black. Then the engraver proceeded to reduce the intensity of the blackness where he wanted lighter areas to appear, by scraping off the roughness with a tool called a rocker. He worked from dark to light, rather than from light to dark, as in the preceding methods. The rocker blade consisted of a number of tiny teeth which made zigzag marks on the plates. It was a laborious task, as the rocker had to be used many times in every direction. When the plate was considered to have been rocked sufficiently, the design was transferred on to it, in reverse, with a scraper.

With this method, as with the aquatint described below, the plates wore each time a print was taken from them, so that no two prints were ever exactly the same. Often the plate needed retouching, which made a further difference to the next print, and on occasion the whole plate had to be re-modelled. The best example of this is *The Temple of Flora*.

Mezzotint plates were often colour-printed, since surface printing methods like this were better suited to colour printing than were line processes. Colour printing itself is described below.

STIPPLE ENGRAVING

This was an etching rather than an engraving technique—in other words acid, rather than a tool, was used to cut into the copper plate. The plate was first covered with a substance such as varnish, which was impervious to acid. Using etching needles and punches the design was then copied over in the form of small dots which penetrated this etching ground and went right through to the copper beneath. The dots were larger and close together where the design was to be the darkest, and small and far apart where a light effect was required. When the acid was applied it ate through the copper in which the dots had been punched—more or less deeply, according to the size of dot—leaving intact only the parts covered with the ground. Very fine details were sometimes added by retouching the plate afterwards. A very delicate and varied tone was produced, which was generally used with colour printing on the plate, if necessary retouched by hand later.

Stipple engraving was the method used for the plates of the great French masters Redouté, Bessa and

PLATE 52
Victorians cultivated a passion for hothouse orchids,
nursing brilliant flowers from dried tropical tubers.
The chromolithographs of Henry Sander's Reichenbachia *(St. Alban's, England 1886–94)*
capture the saturated color of rarities such as
"Cattleya percivaliana," modeled on a watercolor by Henry Moon.

Prévost in the late eighteenth and early nineteenth centuries. Bartolozzi invented the method in England, but it came to full flower in France, and Langlois was perhaps the greatest French engraver.

AQUATINT

The main difference between the aquatint and the stipple engraving was that in the aquatint the etching ground was porous. Acid was applied to the plate many times; as soon as a particular area was sufficiently bitten into, it was covered with a non-porous ground, and so on until the darkest areas had been dealt with. Any area that was to print pure white was completely covered with the non-porous ground to start with. This method produced a kind of soft half-tone effect. As with mezzotints, the plates wore and had to be continually worked on, so apparently identical prints produced by this method may contain a number of minor differences.

SOFT GROUND ETCHING

Soft ground etching was the method used for the plates of John Edwards' *A Collection of Flowers Drawn After Nature*, and for no other natural history book. These plates, which appeared between 1783 and 1801, used 'light' only for the etched tints, and it is possible to think that one is looking at a pure watercolour.

In this method the etching ground was mixed with tallow, and was therefore softer. The design was drawn or traced on a sheet of paper laid over the ground, and the softness of the ground enabled the etcher to cut the lines through it on to the plate beneath. Acid was then applied, and worked in the same way as with stipple engraving.

COLOUR PRINTING

Some natural history plates were entirely colour-printed, while others, including the great works of Thornton, Redouté and Levaillant, were partially colour-printed and retouched by hand afterwards. There were two ways of colour printing from the plate—applying all the colours on to a single copper plate (the method favoured in England), or making several identical copper plates, applying a different colour to each, and printing them on to the paper one after another. This latter method was the one used in France.

LITHOGRAPHY

The technique of lithography was invented accidentally by an Austrian, Alois Senefelder, in 1796. In a hurry one day, he wrote down his mother's laundry list on a handy stone—he had been conducting etching experiments using various objects as 'plates'—and when he later tried to wash off the writing with acid and water he found he could not, since the ink was greasy. By about 1840 lithography had ousted all other techniques.

In the perfected method a smooth lithographic stone received the design in pen and greasy ink or a greasy lithographic crayon. The softness of the crayon gave a rather imprecise line, and lithographic plates were frequently improved by retouching with a scraper or by various other methods. The stone was then wetted, but the inked parts rejected the moisture. When a greasy ink was passed over the stone, however, it was accepted by the previously inked portions. The lithograph was frequently hand-coloured, which accounts for its superiority to the chromolithograph.

Apart from the difference in appearance, a lithographic print does not have the plate mark which surrounds engravings, so is easily distinguishable.

CHROMOLITHOGRAPHY

This is the name given to colour printing using lithographic methods. The plates could be printed many times over, with blocks of colour overlapping and overprinting to create a range of colours. As many lithographic stones were needed as there were colours to be applied. This multiple printing gave the plates a rather greasy, shiny appearance, suitable for the vivid plumage of tropical birds but much less so for flower prints. The process required considerable skill to do it well, because of the multiple printing. This explains why, especially late in the period, so many bad chromolithographic plates were produced.

ABBREVIATIONS

Natural history plates frequently carry a number of credits to the various people involved in preparing them. Since Latin abbreviations are usually used they can be perplexing to the uninitiated. Below is a brief glossary of these terms.

Del.: stands for *delineavit* or *delineaverunt*—Latin for he (or she) or they drew—and therefore refers to the artist. This is usually found on the bottom left-hand side of the print, after the artist's name.

Dir.: stands for *direxit* or *direxerunt*, and refers to the person(s) who supervised the engraving (i.e. not necessarily the actual engraver). The name precedes it.

Exc.: stands for *excudit* or *excuderunt*, meaning (it always follows a name) engraved by, or sometimes engraved and printed.

Fe. or *Fec.*: stands for *fecit* or *fecerunt*, meaning made by. A less common variant of *del.* The name always precedes it.

Imp.: stands for *impressit* or *impresserunt*, meaning printed by. On French plates it often stands for *imprimé* or *Imprimerie*—it all means the same thing. This information is usually found below the title of the plate.

Lith. or *Lit.*: refers to the lithographer, whose name it follows.

Pinx.: stands for *pinxit* or *pinxerunt*, meaning painted by. In this context it is a less common alternative to *del. Pinx.* also sometimes denotes the hand-colourist.

Sc. or *Sculp.*: stands for *sculpsit* or *sculpserunt*, meaning engraved by. It follows the engraver's name and is usually found on the bottom right-hand side of a plate.

BOOK SIZES

The general reader—and the expert will have no need to read this account anyway—wants to know within reason how large a book is, and not to be told this in too technical a way. When a sheet of paper from the printers has been folded once it is called a folio, and produces two leaves, that is four numbered pages. Folded again it is a quarto—four leaves, eight pages. A large number of ordinary books are octavo, with eight leaves and therefore sixteen pages from each original printer's sheet. Smaller books can be 12mo, 16mo, 24mo and 32mo.

The American Century Dictionary gives 30 different sizes of folio, quarto and octavo books. Most readers do not want to know all the details, and do not mind how many times a sheet has been folded. So I have always borne this in mind and written about 'large folio', 'folio', 'small folio', and so on. One problem is that the terms are not used consistently; to me, for instance, a large quarto is a folio, and I call it this! A small quarto is sometimes called an octavo. Quarto books are, however, almost always square or squarish, while folios and octavos are taller than they are wide.

There is one exception to all this—very, very large books, such as Audubon's *The Birds of America*, or the somewhat smaller Thornton's *The Temple of Flora*, must be distinguished somehow, so to me the first is elephant folio (some prefer double elephant folio), and the latter is very large folio. Remember also that most large natural history books came out either in parts, or in their original boards, and the most valuable ones were bound more proudly by their owners afterwards, some in full calf binding, some in half calf, and some in full or half morocco. The best of these bindings are contemporary, but some, less fortunately—for their value at any rate—are modern. In all cases, or very nearly all, this made the original size smaller. If we want to be really fussy, or perhaps it is better to say really accurate, we must give the size of books in inches and centimetres, both in height and breadth. However, since the original sheets from the printers varied in size, this would have to be done for every single book. As an approximate guide we can think of elephant, the largest size encountered, as about 30 x 40 inches (Audubon's *Birds* is actually 29½ × 39½), with folio, quarto and so on correspondingly smaller.

WATERMARKS

Good, hand-made paper such as was used for natural history plates always contained a watermark, made by pressing a wire shape against the paper during the making process. The watermark consisted of the paper-maker's name and usually the date, and was invisible unless held up against the light, when it could, and still can, be seen with a little difficulty. The date is particularly important with mezzotints and aquatints, since every time a print was pressed on to and then pulled off the plate, the plate wore, as described above, so that in fact no two prints were exactly the same.

If a print is dated 1806 and the watermark is dated, say 1818, the print cannot have been a very early issue. If the watermark says 1802, the chances are high that it will be one of the earliest issues. Anyone with a real interest in the subject should look very carefully at watermarks.

THE INDEX